Forest Ecosystem Management and Timber Production

Timber sourcing is shifting from extraction from natural forests to forms of cultivation that are increasingly agricultural in nature. This book takes a multidisciplinary approach to examine the socio-political, biophysical and discursive dimensions of this divergence of wood production from forests. This analysis challenges the historical integration of wood production and forest ecosystem management exemplified by the institutions of forestry with their inherent wood/forest connection. This has significant implications for how wood and forest socio-ecological systems confront change and challenge ideas about how to achieve sustainability.

Historically, the institutions of stewardship forestry were founded on ideals of sustainable systems in long-term equilibrium. However, these occur within rapidly evolving social and technological contexts that constantly challenge the maintenance of any equilibrium. This creates considerable tension within wood and forest socio-ecological systems and their institutions and governance. Moving beyond adaptation to transformation, however, requires a willingness to consider post-forestry conditions, such as integration of emerging wood cultivation systems into agricultural and landscape approaches, and increasing management of extensive forest ecosystems for non-wood values in the absence of wood production. This book includes four case studies: a global modelling of shifts in wood production and three national case studies (Australia, Indonesia and New Zealand), each analysing shifts in resilience in wood and forest socio-ecological systems using a different disciplinary approach.

This book will be of interest to advanced students, researchers and professionals in forestry, land use, conservation, rural studies and geography.

Russell Warman is a University Associate at the School of Technology, Environments and Design, University of Tasmania, Australia. He is a multidisciplinary geographer with over 20 years of professional experience working with communities and landscapes in natural resource management and environmental planning and design.

The Earthscan Forest Library

This series brings together a wide collection of volumes addressing diverse aspects of forests and forestry and draws on a range of disciplinary perspectives. Titles cover the full range of forest science and include: the biology, ecology, biodiversity, restoration, management (including silviculture and timber production), geography and environment (including climate change), socio-economics, anthropology, policy, law and governance. The series aims to demonstrate the important role of forests in nature, peoples' livelihoods and in contributing to broader sustainable development goals. It is aimed at undergraduate and postgraduate students, researchers, professionals, policy-makers and concerned members of civil society.

Recent Titles:

The Decentralization of Forest Governance
Politics, Economics and the Fight for Control of Forests in Indonesian Borneo
Edited by Moira Moeliono, Eva Wollenberg and Godwin Limberg

Forest Ecosystem Management and Timber Production
Divergence and Resource Use Resilience
Russell Warman

Additional information on these and further titles can be found at www.routledge.com/books/series/ECTEFL

Forest Ecosystem Management and Timber Production

Divergence and Resource Use Resilience

Russell Warman

Routledge
Taylor & Francis Group
LONDON AND NEW YORK

earthscan
from Routledge

First published 2019 by Routledge

2 Park Square, Milton Park, Abingdon, Oxfordshire OX14 4RN
52 Vanderbilt Avenue, New York, NY 10017

Routledge is an imprint of the Taylor & Francis Group, an Informa business

First issued in paperback 2020

British Library Cataloguing-in-Publication Data
A catalogue record for this book is available from the British Library

Library of Congress Cataloging-in-Publication Data
Names: Warman, Russell, author.
Title: Forest ecosystem management and timber production : divergence and resource use resilience / Russell Warman.
Description: New York : Routledge, 2019. | Includes bibliographical references and index.
Identifiers: LCCN 2018038175 (print) | LCCN 2018038683 (ebook) | ISBN 9780429485831 (eBook) | ISBN 9781138599239 | ISBN 9781138599239(hardback) | ISBN 9780429485831(ebk)
Subjects: LCSH: Forests and forestry. | Forest management. | Timber. | Forest resilience.
Classification: LCC SD131 (ebook) | LCC SD131 .W37 2019 (print) | DDC 577.3–dc23
LC record available at https://lccn.loc.gov/2018038175

ISBN: 978-1-138-59923-9 (hbk)
ISBN: 978-0-367-66344-5 (pbk)

Typeset in Bembo
by Wearset Ltd, Boldon, Tyne and Wear

Contents

List of figures vii
List of tables ix
Acknowledgements x

1 Wood and forest social-ecological system resilience – setting the scene 1

PART I
Transitions in forest use and wood sourcing 13

2 An environmental history of forest use and wood sourcing 15

3 The evolution of forest use and wood sourcing – a model 46

PART II
Case studies: countries, regions and themes 63

4 Global analysis of trends in wood sourcing 65

5 A natural turn: land use change, leakage and forest conservation 85

6 A social turn: decentralisation in Indonesia – a forestry history 107

7 A discursive turn: foresters' speak, and transitions in New Zealand 128

PART III
Transitions and resilience 151

8 Wood and forests in a post-forestry world 153

9 Resilience, innovation and sustainability in resource use 173

Index 188

Figures

1.1 Continuum of forest and treed landscapes and respective wood sources 5
1.2 Location of three case study countries: Australia, Indonesia and New Zealand 7
1.3 Total industrial roundwood production for cultivated wood sources and natural forest sourced wood for three case study countries separately and combined (1961–2015) 9
2.1 Outline of main forest and forestry environmental history thematics reviewed and the revised thematic used here 17
2.2 World forest extent and industrial roundwood production in 2012 30
3.1 Evolving and diverging basins of attraction in wood production and forest use social-ecological systems showing a temporal sequence from left to right 49
4.1 Continuum of wood source characteristics and the two corresponding sets of cultivated wood used here 68
4.2 Results for total historical industrial roundwood production 1945–1970 74
4.3 Total industrial roundwood production for historical period 1945–2016 and three forecast scenarios for the period 2017–2030 75
4.4 Total fuelwood production for the historical period 1945–2016 and three forecast scenarios for the period 2017–2030 75
4.5 Estimated fuelwood and industrial roundwood production variants from wood cultivation, including trees outside forests 76
4.6 Estimates of global natural forest roundwood production 1945–2030 77
5.1 Basic model of leakage and its relationship to intended outcomes 88
5.2 Matrix of main distinctions of leakage sources and types 89
5.3 Sustainable versus actual yield of sawlogs from Australia's public natural forest 93

5.4 Sawlog and pulplog consumption (ABARES data), housing starts and per capita gross domestic product 94
5.5 Top – Sawlog and pulplog production from natural forest on public and private land in Australia 1996–1997 to 2010–2011. Bottom – Log production from natural forest for the main producing states of Australia 1996–1997 to 2016–2017 95
5.6 Sawlog (top) and pulplog (below) production from natural forest in each Australian state and national production from plantations; total national consumption of logs and national hardwood log consumption 96
5.7 Australian net trade of sawlog equivalents, pulplog equivalents, log equivalents for net furniture trade and total net log equivalents 98
6.1 A schematic of the spheres of interest, including drivers, relationships and the motives of individual spheres 109
6.2 Production of total logs and the three successive waves of log uses 114
7.1 Specificity results of selected words noted in the following discussion 132
8.1 Total global roundwood production (industrial and fuel wood) 1945–2016 and cultivated roundwood and natural forest roundwood portions of total roundwood 153
9.1 Representation of a ship (a social-ecological system) changing direction (transforming) to avoid running ashore on a rocky island (catastrophic collapse of system) 181

Tables

4.1 Summary of total roundwood production data methods and materials 69
4.2 Summary of total cultivated roundwood production data methods and materials 72

Acknowledgements

My biggest thanks goes to Jodie for her uncompromising support – you are a wonder.

Much of the initial work for this book was conducted during my PhD candidature and big thanks go to Pete and all my other colleagues who provided support and camaraderie for this project. I was supported during the research for this book with a Tasmanian Graduate Research Scholarship from the University of Tasmania, and my thanks go to them. Finally, thanks also to Tim, Amy and the team at Routledge.

1 Wood and forest social-ecological system resilience – setting the scene

The rudimentary enterprise of entering a forest, selecting a tree and removing wood has been a staple activity through much of human history. But the sourcing of wood is now undergoing a fundamental shift. From a history as long as humanity's, wood has been collected from natural forest. But increasingly this is shifting to cultivated wood sources such as plantations and trees not from forests. Much as other food and fibre agriculture emerged from earlier methods of collection from wild ecosystems, the same is now happening for wood (Binkley, 2003; Carle *et al.*, 2009; Sedjo and Lyon, 1983). This process began to occur in the last several centuries in a few densely populated parts of the world such as West Europe, Java and Japan. However, the shift to cultivated trees[1] for wood production rapidly gathered pace in the twentieth century (Evans, 2009).

A large body of literature is focused on the environmental, economic and social values and impacts of timber production in both natural forests and plantations. Despite this, knowledge about the process of the transition itself is limited. In part it is speculated here, this is because there has been a tendency for wood production, forestry and forest conservation thinking to be grounded in ideals of sustainability, balance and equilibrium – not dynamic evolution and change. And also because of institutional biases towards the assumption of the natural integration of extensive forest management and wood production. However, this work will demonstrate that this assumption of the wood/forest nexus is being fundamentally challenged. As with any paradigmatic challenge this allows new insight and opens up new opportunities. In addition, there is a tension between goals of sustainability and systemic equilibrium and the siting of such goals in systems of profound social, technological and biophysical change. This work is undertaken with an assumption that better understanding of these processes of change will improve policy in response to evolving natural resource regimes.

At the outset it is worth briefly noting the personal origins of this work in order for you, the reader, to understand where I am 'coming from' in choosing this topic, this approach, this data and the conclusions I reach. In the case of this book much of the research was undertaken as part of my PhD candidature. But the research questions themselves grew from over 20 years work in various forms of natural resource management (including forestry and forest conservation). During these experiences I became interested in understanding how society responds to limits of natural resources through changes in technologies and institutions and the political and economic contestation involved over limited resources and conflicting values. I was also struck by the paradox of systems of resource use displaying seeming opposing qualities of both fragility and resilience in the face of change.

It was from these observations that I launched into the broad research aim of this book: to improve understanding of transitions in resource use regimes. This has been delimited to the case of a single global resource use system here – wood and forests. And specifically, an examination of the global transition of wood production, from extraction of wood from natural forest to cultivated wood sources. The inquiry asks three questions related to this transition: *Why is the transition occurring*? In particular, what are the changing historical, environmental, social, political and economic factors driving this? *To what extent is the transition occurring*? Including, to what extent is it likely to occur into the future? And finally, *what are the implications of this transition?* This last question has obvious relevance for systems of wood production and forest use. But, it is also pertinent to other natural resource use transitions.

These research questions are approached with an open attitude to discipline. This book could be said to be a political economy analysis in the traditional use of the term. It is *economic*, in the sense of pursuing questions of how people maximise the utility of limited resources (wood and forests), and *political*, in the sense of pursuing questions of how people utilise and seek power over access and use of limited resources. In this multidisciplinary spirit the research here delves into economics, politics, sociology, geography, ecology, conservation biology and, of course, forestry. It eschews the institutional reflex to delimit the research project by focusing on a singular approach, discipline or set of data. Instead, it seeks what can be illuminated when the phenomenon being considered and critiqued is approached as a woven network viewed from multiple directions. Of course, this still requires delimitation, undertaken as set out in the following description. But it allows an immersive approach into the networked nature of the phenomena of evolving forest and wood production systems.

Three key areas of theory are used to frame this work. First, actor network theory informs a multidisciplinary approach to complex systems being studied. Second, resilience thinking is drawn upon to describe the complex systems being studied and how they change. Finally, particularly in Part III, this work draws on the theory of reflexive modernity to consider how these complex systems, and the contemplation of them, are shaped.

The actor-network theory's distinctive approach to treating the social, discursive and natural in complex networks (Law, 1992) provides a helpful framework for undertaking multidisciplinary analysis of complex adaptive systems. The limits of singular approaches that actor-network theory seeks to address are noted in this paraphrased quote from Bruno Latour:

> [Forestry is] too social and too narrated to be truly natural; the strategy of [foresters, timber companies and forestry bureaucracies] too full of [yield analyses and forest mensuration] to be reduced to power and interest; the discourse of [forests and wood] is too real and too social to boil down to meaning effects.
>
> (Latour, 1993, p. 6)[2]

For the study of wood production and forests this approach provides a useful conceptual basis for the multidisciplinary and grounded approach that is taken. It recognises the multi-faceted nature of the problem and the phenomenon. The topic of study is approached from multiple directions. None alone reveals an understanding sufficiently removed from the subject, but, taken together, there is the potential for a deep analytical penetration.

Resilience thinking (Holling, 1973, 2001) provides a useful conceptualisation for exploring the formation and evolution of forest use and wood production social-ecological systems (hereafter, *wood and forest social-ecological systems*). It considers how these social-ecological systems develop certain configurations (a *state* or *basin* to use the language of resilience thinking) that give the system resilience in the face of pressures on system conditions. Further, this conceptualisation supports analysis of how systems maintain resilience in the face of change, but also transform or transition into new states. Resilience thinking lends itself to considering whole systems and how they move from one state or condition to another. Resilience thinking also provides a place where the path-dependent nature of institutional and technological processes can be considered in relation to social systems, as it can explain systematic resilience and transformative capacity (or fragility). Resilience theory, too, has come to better incorporate the feedback complexities of the social actors in its social ecological systems. Together, both actor network theory (with origins in social sciences) and resilience thinking (with origins in natural sciences) provide useful ways of thinking about complex social/natural systems such as the wood and forest social-ecological system.

Finally, reflexive modernity (Beck, 1992; Beck *et al.*, 1994; Giddens, 1990) also provides theoretical framing for understanding and thinking about the social institutions being studied here. The central idea of this theory is that a contemporary modernity has emerged that reflects on the challenges and problems (*risks*) of key features of modernity, such as rapid technological change, capitalist and industrialist economies, a pre-eminence of rationality and reason, and the nation state. These features of modernity will play an important role in describing the evolution of wood and forest social-ecological systems. Related to this, reflexive modernity within the 'risk society' points to the need for resolution of emerging conflict about how to resolve tensions within social-ecological systems through risk negotiation. Appeals to higher authorities (including an objectively detached science) are limited in their ability to reconcile the normative tensions that exist in the systems being studied. Reflexive modernity also serves as a reminder that this work itself will be an act of reflexive modernisation – it is not only dealing with problems of modernity, but the work itself exists within and affects the system being studied. The work here recognises that the making of choices in an environment of political contestation and negotiation is key to resolving tensions between understandings and values acting on the social-ecological systems of study.

For this book there is a need to address part of this contestation up front. At the heart of this book's topic are the distinctions between various wood sources, ranging from natural 'virgin' forests to industrial plantations and trees not from forests. Forests and trees (as wood sources) occur within a broad spectrum of characteristics across the globe. Furthermore, many forest definitions have been established within institutional frameworks focused on wood production, but are found to be wanting when researchers and policy makers direct attention to forest restoration, conservation, agroforestry and management of mosaic landscapes (Chazdon *et al.*, 2016). This presents real definitional challenges. To address this there is a need for creating definitions that are fit-for-purpose (Batra and Pirard, 2015; Chazdon *et al.*, 2016). This is particularly true for this book – there needs to be a clear distinction between *natural* and *non-natural* wood sources.

What is meant by *natural*? Joseph Pitt (2011) observes that the terms *natural* and *artificial* are problematic, with the distinction itself 'contrived'. Much that we label *natural* is influenced in some way by human agency, while the artifices of humanity have entirely arisen from within nature. Pitt (2011, p. 82) suggests that 'we should finally give up that worn out old distinction between the natural and the artificial'. Hodgson (1993. p. 34) concurs, noting that 'the Cartesian and Newtonian worldviews have sanctioned habits of thought which involve an ultimately untenable conceptual division between humankind and the remainder of the natural world'. Yet forests and wood sources differ, and these distinctions are useful and important. Central to the questions of this book are the *natural* or *cultural* qualities of forests and wood sources.

For labelling different forests, the conceptual opposite to natural, *cultural* (or *artificial*), is not generally used, but rather *plantation*, or more recently *planted forest*, referring to a planting of areas of trees for wood production or other purposes. However, as will be explained in more detail later, many forests that are considered to be natural have had significant human influence in their development. Fire, as well other wild food collection and hunting support processes of forest disturbance – undertaken to encourage preferred species for food and other uses – have all influenced the nature and structure of forests. Because of this, the present condition of those forests cannot easily be said to be purely *natural* (in the sense of being somehow free of human affects).

However, there are qualities to forests (or other assemblages of trees) that can be distinguished as *natural* that describe key features which different forests have to a lesser or greater extent. For this work, the key quality defining naturalness is the presence of an ecosystem containing a complex suite of species that are largely indigenous to the area and whose presence in that locale can be largely attributed to past processes other than human intervention. For further refinement it might also be useful to consider that the trees that dominate the forest are primarily the result of natural regeneration, or that human processes of disturbance have nevertheless maintained complex forest ecosystems with their suite of primarily indigenous species. In this work these forests will be referred to as *natural forests* or *forest ecosystems*. *Natural forest* will be used in contexts where the relevant literature uses this term, while *forest ecosystems* will be used elsewhere. Whichever term is used these are distinct from other assemblages of trees that lack these qualities of being complex ecosystems with indigenous tree species predominating. These other assemblages of trees include plantations and many trees grown in rural and urban landscapes.

There are of course a range of conditions that occur in between natural forests and plantations and other plantings of trees. This problem has had some attention in the past decade. Carle and Holmgren (2003) undertook a review of definitions employed in international forest and forest product data collection and analysis. This led to a spectrum description based on naturalness and management intensity in forests. They used the term *planted forest*, which includes *plantation*, as well as the more intensively planted subset of *semi-natural* forests called *planted semi-natural* (see also Kanninen, 2010; Penna, 2010). The reality is that wood sources occur on a spectrum with no neat dividing line, especially when analysed at the global scale. The definitions developed (Figure 1.1) have been adopted in international forestry discourse and will be useful in this work.

<table>
<tr><th colspan="4">Natural forest wood</th><th colspan="2">Cultivated Wood</th></tr>
<tr><td colspan="2">Natural

Essentially naturally regenerated</td><td colspan="2">Semi-natural

Intensive silvicultural assistance in regeneration of primarily native species</td><td rowspan="2">Plantation

Primarily introduced species</td><td rowspan="2">Trees outside forests</td></tr>
<tr><td>Primary</td><td>Modified</td><td>Assisted natural regeneration</td><td>Planted</td></tr>
<tr><td>No clearly visible signs of human activity</td><td>Clearly visible signs of human activity</td><td>Activities such as weeding, fertilizing, thinning and selective logging</td><td>As for 'Assisted', but also includes planting, seeding and coppicing of specific species</td><td>Planting of species for afforestation or reforestation, often as monoculture</td><td>Stands smaller than 0.5 ha in agricultural and urban landscapes</td></tr>
<tr><th colspan="4">Forest</th><th colspan="2">Non-Forest</th></tr>
</table>

Figure 1.1 Continuum of forest and treed landscapes and respective wood sources.
Source: modified from Carle and Holmgren, 2003.

Using the above spectrum allows a more careful siting of the basic duality being examined here – shifts in wood sourcing between natural forests and cultivated trees. A distinction is made between two main categories of wood sources. These are *natural forest wood*, as defined already, and *cultivated wood*. The latter term is used to include plantations, agroforestry and rural and urban tree plantings – that is, trees planted through active tree cultivation for productive purposes. The term *cultivated wood* expands on the widely used dichotomy of *natural forest* and *plantation* (for example Ajani, 2011; Brown, 2000; Sedjo, 2001) by including the role of trees outside forests (Food and Agriculture Organisation of the United Nations 2010 [FAO]).

The definition of *forest* itself is fraught. However, using the term *wood cultivation* for the activity of growing the wood and *cultivated wood* for the resultant wood, allows some circumvention of the debate about whether trees grown using largely agricultural models of production (particularly plantations) should be considered as forests (Sasaki and Putz, 2009; World Rainforest Movement, 2003). It recognises that 'trees and forests are not synonymous' (Long and Nair, 1999, p. 146), and that trees outside forests are a growing and significant source of wood.

Distinguishing between natural forests and wood cultivation starts to help in considering change in wood and forest social-ecological systems. *Cultivated wood* might better sit in agriculture rather than forestry. This research will generally refer to the activity of *wood production* (covering the full range of wood sources) or *wood cultivation* (covering wood sources not from natural forests), rather than *forestry* when the discussion is focused on generating wood. *Forestry*, will be used to describe the institutions[3] that have developed around the stewardship of forests for wood production. *Wood cultivation* and *cultivated wood* might more clearly describe the production of wood than *forestry* as *forestry* runs into some limitations in relation to wood production from sources that are outside *forests* and when forests are managed for purposes other than for wood production such as in protected area management.

After this introduction the book is structured into three parts. Part I provides an overview of the transition of wood production and is conducted by way of an environmental history that captures the substantive phases and qualities of these changes. It focuses on the broad sweep of human use of wood and forests and the changing interconnections between the two. Part I, then, proposes a heuristically derived model of the transition using resilience thinking. This model is used to describe the sequence of different states for wood production and forest use social and ecological systems. The three sequential states are: forest exploitation, stewardship forestry and then two simultaneous, but geographically divergent, systems of forest ecosystem management and wood cultivation. The key feature of this divergence is the undoing of the logic of the wood/forest nexus that underpins forest exploitation and particularly stewardship forestry.

Whereas Part I reviews the historical nature of change in wood and forest social-ecological systems, Part II looks at this shift through a global analysis and then three national case studies: Australia, Indonesia and New Zealand (see Figure 1.2). It starts at the global level and undertakes a quantitative assessment of the transition of wood production from natural forest exploitation to cultivated wood sources. Through a model based on international datasets this first chapter answers the question of the extent of the transition of wood production from natural forest exploitation and management to wood cultivation. Further analysis of transitions in wood and forest social-ecological systems is then triangulated through three national case studies with a mix of methods used. The national case studies tackle the phenomenon from three different approaches to understanding complex networked systems, such as wood and forest social-ecological systems.

Bruno Latour notes the 'three tacks' critics can take when 'talking about our world': naturalisation, socialisation and deconstruction. He says there is a 'common sense' to this 'critical tripartition' that helps illuminate how networks of things and relationships are '*simultaneously real, like nature, narrated, like discourse, and collective, like society*' (Latour, 1993, p. 6 [author's italics]). Australia is examined by looking at change in forest use patterns from a land use science and economics perspective. Indonesia is approached from a socio-political perspective looking at the history of institutional change in forestry and shifts in related power relations. Then, New Zealand is considered through a discourse analysis of the writings of foresters over a nine decade period in which wood and forest use systems underwent major changes. While different tacks in each case study limit the scope for comparison, it does provide a deeper understanding of the phenomena being discussed.

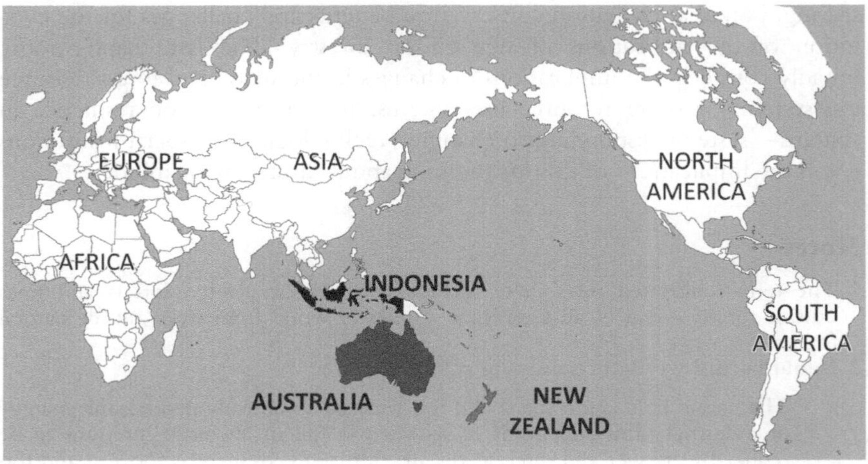

Figure 1.2 Location of three case study countries: Australia, Indonesia and New Zealand.[4]

The three case study countries all have significant areas of forest and wood product industries. Australia and Indonesia are ranked seventh and eighth respectively in the world for areas of forest. Between the three countries they have 226 million hectares of the world's forests – over 5 per cent (FAO, 2015) – ranging from cool temperate to equatorial tropics regions. In 2014 they produced 6 per cent of the world's industrial roundwood. The international rankings for industrial roundwood production include: Indonesia seventh, New Zealand thirteenth and Australia fifteenth (FAOSTAT, 2015). Figure 1.3 below shows the industrial roundwood production for the three countries over the period 1960–2015 and the portions of plantation and natural forest sourced wood produced.[5]

While these three countries are found in one region (Figure 1.2) they are significantly different and thus more broadly representative of processes of change across the globe. They include both developed and developing countries at different stages of the transition being described. New Zealand is a country whose wood production is almost entirely based on cultivated wood from plantations (see Figure 1.3). Meanwhile, Australia and Indonesia are each less 'advanced' along this path, but their patterns of wood production show they are shifting in the direction of New Zealand. These latter two countries also contain a number of sub-jurisdictions that are at different stages of the transition with each encompassing a diverse set of political, cultural and biophysical conditions. For example, the Australian state of South Australia, with a small natural forest estate, has a long established plantation industry which is very different to the state of Tasmania that, until recently, has generated the majority of its wood from a rich natural forest endowment. Together these three countries cover a broad range of conditions.

Finally, Part III then considers potential contributions and conclusions of this work. It does this in two chapters. The first focuses on the wood and forest social-ecological system and particularly current issues of land sharing/sparing, potential plantation conservation benefits and challenges for the evolution of the institutions of stewardship forestry. The last chapter more broadly considers the implications of changes in the wood and forest resource use system for other resource use systems. It concentrates on resilience in systems – systems responding to changing technologies and social institutions – and the implications of this for thinking about future sustainabilities.

Notes

1 The term 'cultivated trees' here includes plantations, planted forests and trees outside forests – that is, all significant sources of wood from trees outside natural forests.
2 Latour's actual words (in translation at least) were:

> The ozone hole is too social and too narrated to be truly natural; the strategy of industrial firms and heads of state is too full of chemical reactions to be reduced to power and interest; the discourse of ecosphere is too real and too social to boil down to meaning effects.

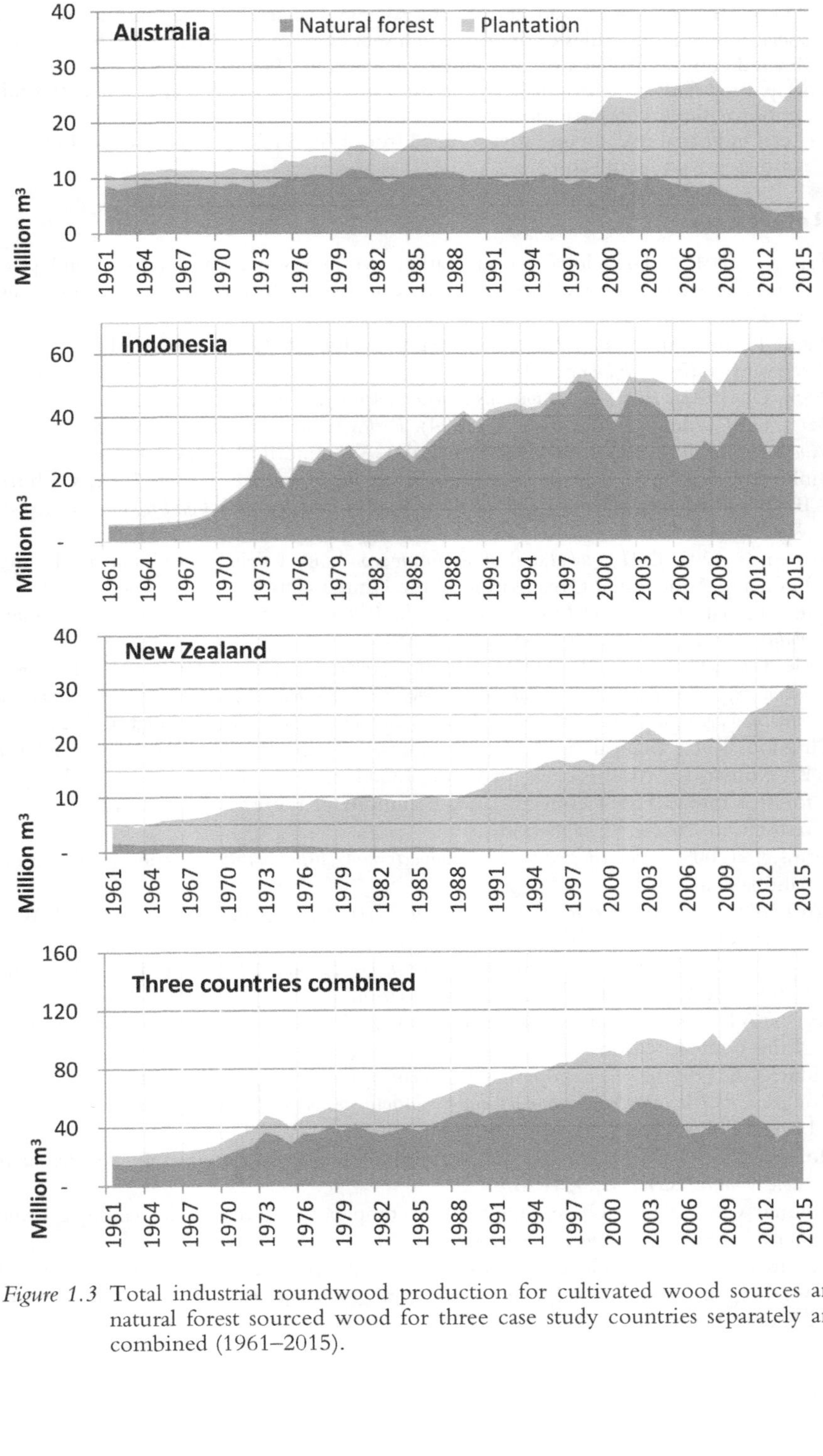

Figure 1.3 Total industrial roundwood production for cultivated wood sources and natural forest sourced wood for three case study countries separately and combined (1961–2015).

3 The term *institution* is used here in the sociological sense – the cultural, political, social or economic 'rules of the game of a society or more formally … the humanly devised constraints that structure human interaction' (North, 1995, p. 3).
4 Map modified from original; by permission from: www.d-maps.com/m/world/centrepacifique/centrepacifique14.ai.
5 Total industrial roundwood data sourced from FAOSTAT and plantation data from national government sources.

References

Ajani, J. (2011). The global wood market, wood resource productivity and price trends: an examination with special attention to China. *Environmental Conservation*, 38, 53–63.

Batra, P. and Pirard, R. (2015). Is a typology for planted forests feasible, or even relevant? *CIFOR Infobrief*, 121, 1–8.

Beck, U. (1992). *Risk society: towards a new modernity*. London: Sage Publications.

Beck, U., Giddens, A. and Lash, S. (1994). *Reflexive modernization: politics, tradition and aesthetics in the modern social order*. Stanford, CA: Stanford University Press.

Binkley, C.S. (2003). *Forestry in the long sweep of history*. In: Teeter, L.D., Cashore, B.W. and Zhang, D. (eds.) *Forest policy for private forestry: global and regional challenges*. Wallingford: CABI.

Brown, C. (2000). *The global outlook for future wood supply from forest plantations*. Rome: Food and Agricultural Organisation of the United Nations.

Carle, J., Ball, J. and Del Lungo, A. (2009). *Planted forests – uses, impacts and sustainability*. Rome: CAB International and FAO.

Carle, J. and Holmgren, P. (2003). Definitions related to planted forests. *UNFF intersessional expert meeting international steering group on 'the role of planted forests in sustainable forest management. Maximising planted forests' contribution to SFM'*. Wellington, NZ.

Chazdon, R.L., Brancalion, P.H., Laestadius, L., Bennett-Curry, A., Buckingham, K., Kumar, C., Moll-Rocek, J., Vieira, I.C.G. and Wilson, S.J. (2016). When is a forest a forest? Forest concepts and definitions in the era of forest and landscape restoration. *Ambio*, 45, 538–550.

Evans, J. (2009). *Planted forests: uses, impacts and sustainability*. Rome: CABI International and FAO.

FAO (2010). *Global forest resources assessment 2010: main report*. Rome: Food and Agriculture Organization of the United Nations.

FAO (2015). *Global forest resources assessment 2015: desk reference*. Rome: Food and Agriculture Organization of the United Nations.

FAOSTAT (2015). *Statistical databases*. Rome: Food and Agriculture Organization of the United Nations.

Giddens, A. (1990). *The consequences of modernity*. Cambridge: Polity Press.

Hodgson, G.M. (1993). *Economics and evolution: bringing life back into economics*. Cambridge: Polity Press.

Holling, C.S. (1973). Resilience and stability of ecological systems. *Annual Review of Ecology and Systematics*, 4, 1–23.

Holling, C.S. (2001). Understanding the complexity of economic, ecological and social systems. *Ecosystems*, 4, 390–405.

Kanninen, M. (2010). Plantation forests: global perspectives. In: Bauhus, J., Van der Meer, P. and Kanninen, M. (eds.) *Ecosystem goods and services from plantation forests*. London: Earthscan.

Latour, B. (1993). *We have never been modern*. Cambridge, MA: Harvard University Press.

Law, J. (1992). Notes on the theory of the actor-network: ordering, strategy, and heterogeneity. *Systems Practice*, 5, 379–393.

Long, A. and Nair, P.K.R. (1999). Trees outside forests: agro-, community and urban forestry. In: Boyle, J.R., Winjum, J.K., Kavanagh, K. and Jensen, E.C. (eds.) *Planted forests: contributions to the quest for sustainable societies*. Dordrecht, The Netherlands.

North, D.C. (1995). The new institutional economics and third world development. In: Harriss, J., Hunter, J. and Lewis, C.M. (eds.) *The new institutional economics and third world development*. London: Routledge.

Penna, I. (2010). Understanding the FAO's wood supply from planted forests' projections. Ballarat, Australia: Centre for Environmental Management, University of Ballarat.

Pitt, J.C. (2011). Working the natural/artificial distinction. *Philosophy of engineering and technology*, 3 (2), 73–83.

Sasaki, N. and Putz, F.E (2009). Critical need for new definitions of 'forest' and 'forest degradation' in global climate change agreements. *Conservation Letters*, 2, 226–232.

Sedjo, R. (2001). The economic contribution of biotechnology and forest plantations in global wood supply and forest conservation. In: Strauss, S.H. and Bradshaw, H.D. (eds.) *Proceedings of the First International Symposium on Ecological and Societal Aspects of Transgenic Plantations*. Corvallis, Oregon: College of Forestry, Oregon State University.

Sedjo, R.A. and Lyon, K.S. (1983). Long-term forest resources trade, global timber supply, and intertemporal comparative advantage. *American Journal of Agricultural Economics*, 65, 1010–1016.

World Rainforest Movement (2003). *Plantations are not forests*. Montevideo: World Rainforest Movement.

Part I

Transitions in forest use and wood sourcing

2 An environmental history of forest use and wood sourcing

The drive to use forests for wood has been important in the ebb and flow of humanity, and forests across the planet. Our relationship with forests and wood is woven inseparably into the natural world within which we exist. The natural resource endowment of standing forests that many societies have used for wood has been central to the evolving political economy of wood production. As processes of exploitation, expansion and conquest reach their limits and endowment forests become exhausted, alternatives have been sought – particularly through stewardship of the remaining forests and cultivation of trees. As this broad process of change is considered here, a global environmental history of wood and forests is mandated.

First, this requires a suitable thematic structure. Four approaches are reviewed briefly. They are not comprehensive but capture a diversity of ways of framing history. They provide some guidance as to a suitable thematic for structuring an environmental history of wood and its sources. As a start, in structuring his environmental history, Simmons (2008) uses the theme of energy. He identifies four broad historical phases around primary energy sources: the gatherer-hunter[1] phase: the more intensive collection of solar energy use through cultivated crops, pasture, wind and wood fuel starting around 10,000 years ago; the period from 1750 to 1950 in which an industrial society dominated through the opportunities supplied by fossil fuels; and finally, the period 1950 to the end of the millennium in which fossil fuels, and also other energy forms, combined with rapidly expanding global information exchanges to create the post-industrial era. This approach has particular relevance to an environmental history of wood, as a significant portion of historical wood use has been for energy.

For a different approach, Hyde (2012) uses the role of marginal economic activity to describe a Von Thunenian[2] model of change in wood production from forests. Hyde delineates three stages of forest use. The first is the New Forest Frontier, this being wood production conducted as a land clearing activity to create agricultural land, with the wood itself as a free endowment. This stage has two distinct subsets, one being that of shifting cultivators moving through the landscape on a cyclic pattern and the other being permanent settlements with a shifting frontier. The second stage, Developing Forest Frontier, occurs when the marginal cost of clearing more agricultural land becomes higher than the benefits. However, demand for wood is such that the frontier extends now by the logging of forests, without clearing the land as an objective. Whereas agriculture led wood extraction in the previous stage, now agricultural expansion might (or might not) follow the expansion of logging into the frontier. The third stage, the Mature Forest Frontier, occurs when the marginal cost of producing wood through sustainable forest management, plantation establishment and trees outside forests that are located closer to the centres of settlement equals or is less than the cost of exploiting the receding natural forest frontier. This approach is useful as it centres on key economic drivers for change in wood and forest social-ecological systems. It also describes the central transition of wood production to wood cultivation.

Finally, two historical case study approaches to the evolution of forestry are considered. Lane and McDonald (2002) hypothesise stages in the evolution of forest management, largely based on Australian, Canadian and United States analysis. They suggest five stages: '(i) traditional gatherer-hunter society, (ii) exploitive colonisation, settlement and commercialisation, (iii) wood resource protection, (iv) multiple-use management, and finally, (v) sustainable forest management or ecosystem management'. They propose these stages as a 'general model of forest management through time' (2002, p. 193). In a similar vein, Kimmins (2002, p. 265) outlines five phases in the evolution of forestry: a pre-industrial low level sustainable exploitation; a non-sustainable exploitation phase as populations grow; the commencement of 'administrative forestry'; an ecologically based forestry (wood biased, however); and then, 'social forestry – ecologically based, multi-value ecosystem management'. A limitation of these two approaches is that they are both grounded in the discipline of forestry and the logic of wood production as an integral part of forest ecosystem management. They seek to explain how people (including foresters) have responded to changing demands for wood production and its conflict with other emerging demands from forests (such as nature conservation). These themes are limited by their adherence to the logic of the institutions of forestry and the wood/forest nexus. For the purposes of this chapter, then, a thematic structure is used that hybridises the above and incorporates a parallel post-forestry thematic. The six themes proposed are:

1 a Forest utilisation – Pre-agricultural
  b Forest utilisation – Agricultural
2 a Stewardship forestry – Wood primacy
  b Stewardship forestry – Balance and multiple use
3 a Post-forestry – Wood cultivation
  b Post-forestry – Ecosystem service provision

The first stage of *forest utilisation* relates to the opportunistic utilisation of endowment forests arising from their presence in the home range of pre-agricultural peoples and expanding agricultural populations. The second stage of *stewardship forestry* involves the emergence of institutions to manage forests and their wood production – the quest for sustainability. The last stage is labelled *post-forestry*. It is distinguished from the previous two stages in that its two subsets are spatially rather than temporally distinguished. That is, they occur simultaneously in separate locations rather than sequentially in the same location. Figure 2.1 gives an outline of the four different approaches and the synthesised thematic outline to be used. Note that while they are presented as distinctive periods to delineate a workable thematic their sequencing in reality was obviously more intergraded. Likewise, the relationship of the other thematics illustrated are generalised and indicative of stages that will overlap and intergrade to various degrees.

Forest utilisation – pre-agricultural

Most of the natural forests of the world have a history of human interaction. The advent of agriculture, or the arrival of 'first contact' colonising modern societies, was not the beginning of human interaction and alteration of forests (Simmons, 2008, p. 45). Prior changes wrought by humans included modifying forests with disturbance to encourage certain plant and animal species

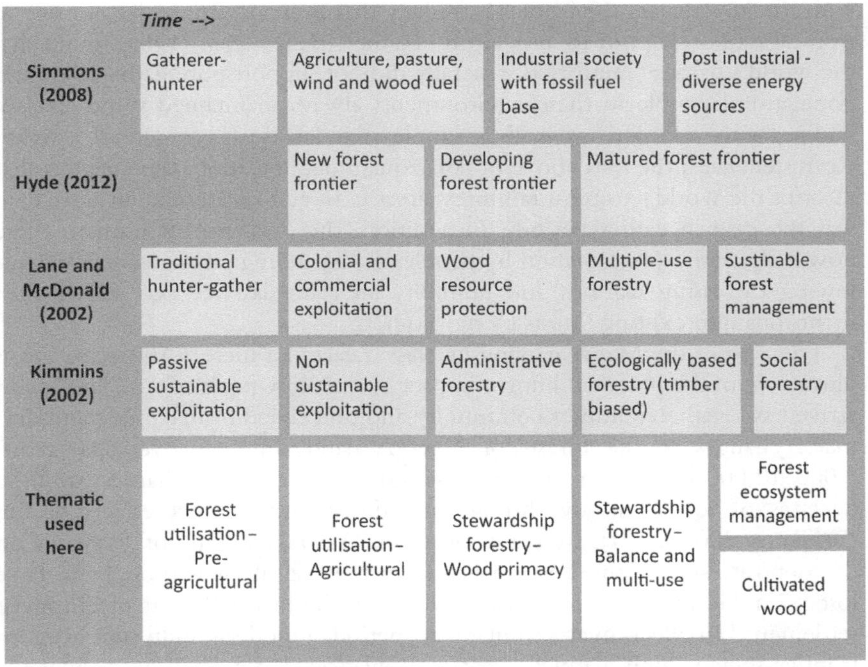

Figure 2.1 Outline of main forest and forestry environmental history thematics reviewed and the revised thematic used here.

(Hladik *et al.*, 1993) and extensive use of fire to modify forests and other landscape ecosystems (Bowman *et al.*, 2011; Garcin *et al.*, 2018; Simmons, 2008). Indeed, it is possible that early phases of human-created deforestation altered not just the places within the widespread footprint of hominids, but the carbon cycle of the planet itself (Ruddiman, 2012).

Swidden[3] styles of agriculture have been considered to occupy a transition point between gatherer-hunter and agricultural phases (for example, Bailey *et al.*, 1989). This form of land management, at the interface of forests and agricultural lands, has been widely practiced globally and has been a significant form of food security (Parrotta *et al.*, 2015). The clearing of small areas in forests has occurred on a spectrum, from small clearings to encourage preferred browsing fauna and pioneering flora to clearing for deliberately planted crops and holding of domesticated animals. Importantly, Bailey *et al.* (1989) point to the early emergence of competition between forests as sources of wood and wild foods on the one hand, and the use of the land cleared of forest for agriculture and settlement on the other. As an example, the extensive tropical forests of the Amazon basin are the product of an extended period of human interactions, both gatherer-hunter and agricultural (for example, Piperno *et al.*, 2017), such that 'the possibility exists that the Amazon of today's environmental concern is mostly a product of 300 years of depopulation since European incursions.' (Simmons, 2008, p. 34) This aligns with the idea that in the tropics it is likely that humans simply did not live in tropical rainforest prior to agriculture. Rather, Bailey *et al.* (1989) found that the world's tropical rainforests generally did not support pure gatherer-hunter populations; people in these environments always maintained some level of trading with outsiders, as well as simple swidden-type agricultural activity. Consistent with the idea above of Amazonian depopulation, this suggests that most of the world's tropical rainforest areas exist at a greater extent now than was the case in earlier periods. In addition, the existence of humans there now is the result of adaptation by peoples to expanding rainforest regions following the rising warmth and humidity after the last ice age, rather than incursions into existing forests by newcomers.

The case of the Maori arriving in New Zealand a mere 1,100 years ago is illustrative of the potential human impact even at low populations. The Maori arrived as a gatherer-hunter community and proceeded to make vast and dramatic changes to the forests of New Zealand. In little over 200 years, 8,000–12,000 people on the South Island destroyed more than 3.2 million hectares of forest, mostly through the direct and indirect effects of fire (Williams, 2002, p. 21). By comparison, the Australian island of Tasmania on the opposite side of the Tasman Sea from New Zealand is thought to have lost 1.7 million hectares of its natural forest in the first 200 years of European settlement.[4] In effect, over a similar time period, a modern, industrial, western frontier society with a much greater population density had a considerably smaller impact on its forests than that of the much smaller group of gatherer-hunters in New Zealand.

The impact of changing global climate conditions sometimes confounds evidence for changes in forests wrought by gatherer-hunter peoples and emerging agricultural societies. In particular, away from the tropics, the ending of the last ice age 12,000 years ago, with the recession of the ice sheets and warming climates, saw the march of forests north across the higher latitudes of the northern hemisphere (Williams, 2002). This occurred simultaneously with the shift of humans into these areas making it likely that the new forests emerged in concert with human influences, but also making the extent of human influence difficult to decisively determine given its occurrence in the depths of prehistory.

What is clear is the world's current stands of natural forests are significantly influenced by an extensive history of human interaction – Simmons refers to the world's 'near-natural forests' (2008, p. 34). This period of human history is a reminder that while the differences between natural forests and non-natural planted forests might seem immediately obvious, the assumptions behind that difference demands considered thought.

Forest utilisation – agricultural

Wood, and the forests that provide it, have been a significant historical driver in the waxing and waning of societies, and the struggles between them. As noted above, it is not easy to disassemble the exact influence of humans in shaping the world's extant forests, but it is clear that they did to some degree and that this was also likely entwined with the evolution of agricultural practices and agricultural societies. Critically, wood was through all this time the primary fuel source for human societies. It was also often a critical material for shelter and tools. Another feature of wood use in this period was its 'pre-forestry' nature, wherein the use of wood was 'simply the use of a forest resource without any explicit management activity to promote or ensure the renewal of desired values' (Kimmins, 2002, p. 264). It was the earliest and most explicit use of naturally generated forests as endowment forests.

It was also during this time that wood, which was not known to have been a particularly limiting resource for previous populations (food generally was much more limiting), became a more critical resource. Suddenly agricultural societies, with their greatly expanded food production and populations, began to confront localised limits to wood supplies. As populations grew in size with the benefits of agricultural food production, the resulting overexploitation of wood resources at times contributed to dramatic social changes, including subsequent expansionary warfare to control wood resources further afield (Simmons, 2008; Vogt *et al*., 2007).

As well as this demand for wood itself there was also the demand for the land on which forests stood. For example, in the classical period forest use for wood probably played second fiddle to the impact of agricultural clearance on forests (Williams, 2002, pp. 79–80). Williams also ventures that the impact of clearance and subsequent agriculture and use of domesticated stock had a much greater impact on the forests than the writings of the time indicate, with their emphasis on the impacts of ship building, metal smelting, supplying fuel and providing construction wood for new and growing cities. He argues that this is a result of forest impacts being subsumed into the tasks of agriculture and often being the work of slaves – thereby getting little attention in the historical record. Critically for forests over the next few millennia was the emergence of maritime trade and its demand for wood to build boats (Williams, 2002). The development of silver for currency, the use of slave labour and the joining together of disparate regions by the ability to transport bulky materials over large distances by ship spawned an emergent pattern of maritime trade – a harbinger of the global blossoming of the use of ships and seas to connect agricultural societies over the next two millennia.

While we briefly consider forest utilisation during the period of emergent agricultural societies here, a question arises – why did the shift from gatherer-hunting to agriculture occur such a long time ago in human history, whilst the shift in wood collection from natural forest exploitation to wood cultivation is much more recent, indeed still underway? To answer the question of the late change to wood cultivation, Sutton (1999) conducts a 'back of an envelope' estimate. He estimates that for a gatherer-hunter about 100 hectares per person is needed for food versus 0.2 of a hectare per person for wood. This created pressures and incentives to develop agriculture much earlier than the cultivation of trees for wood. The far greater endowment of wild systems to meet wood needs meant that the incentives to develop wood cultivation came much later than they did for terrestrial food sources. Interestingly, fisheries, like wood, are only now witnessing the transition away from wild caught sources (Belton and Thilsted, 2014).

Stewardship forestry – wood resource sustainability

What is meant by *stewardship forestry*? Stewardship forestry emerges when it is recognised that extant forests are being utilised at a rate that is unsustainable. In response forms of management and restriction are established to manage the wood resource for a more sustainable supply. It was in early modern Europe that the principles of stewardship forestry that would be carried forward into the world system emerged, however stewardship forestry in the broad sense of the terms also emerged elsewhere, and in earlier times.

Historically, in various places there has been the development of taboos and cultural practices that limited or controlled the taking of certain wood or the disturbance of particular trees, groves or forests (Salick *et al.*, 2007). During the Chou Empire of China (*c.*1100–256 BC) the Emperor dictated that a forest service be established to look after existing forests and undertake reforestation (Hermann, 1976). In Japanese society intense demand for wood in construction and industry from at least the seventeenth century led to various forms of stewardship forestry in response to overharvesting (Iwamoto, 2002). In medieval Europe, the use of forests for wood was largely approached through systems of rights assigned by royalty and the church that allowed access for procurement of wood and related materials. Extracting resources was widely allowed under these rights, except for hunting which the nobility kept to themselves (Williams, 2002). From this approach grew some of the earliest records of forest regulation – regulation that grew from an emerging sense that the forests were limited and had values worthy of protection. Population increases, along with an increasing concern that the 'frontier' could not expand into the forests forever, led to royal charters and ordinances in Europe from the late thirteenth to the middle fourteenth centuries (Williams, 2002, pp. 130–137). Ironically, this period of concern about forest over-exploitation was ended by the plague, which killed more than a third of the population of Western Europe over several years around 1350. There was also a dramatic retreat of population from the margins due to weather deterioration and the war and pillage that followed into the fifteenth century – '[f]or a while, the forest had been reprieved'. (Williams, 2002, p. 136)

Eventually however, the European stewardship forestry approach, which would spread across the globe, emerged with the recognition of a growing crisis in forest depletion. This was brought on by the large volumes of wood used for fuel and the growth in wood demand for building and shipping. It led, for example, to the enactment of laws such as the French Forest Ordinance of 1669 (Hughes, 2006, p. 107). Similarly, in the eighteenth century in Germany forest loss became increasingly apparent and a government concern. This was a time when the modern profession of forester emerged and forest management specifically for wood production started. Wilhelm Gottfried von Moser published the influential text *Principles of Forest-Economy* in 1757. In the European case these early approaches to forest management in Europe were often more akin to the forest management approach known as 'plantations' in the twentieth century, with their emphasis on standard species, planted rows and sustainable yields from monocultures (Simmons, 2008). However, this technocratic forest management, steeped in the rational thinking of modernity, would spread across the globe and be adapted and applied to the extensive natural forest ecosystems of the colonial periphery. This is a critical point for this work, focused as it is on the shift of wood production from natural forests to cultivated wood sources. In a sense, the modern stewardship forestry institutions were often first based on cultivated wood sources – the replanted forests of Europe. Only after were they then applied more extensively to the natural forests of the colonised periphery.

The forestry that emerged in Germany had strong patriotic aspects to its practice (Schama, 1995), spreading as it did with the rise of the notion of the nation state. This was consistent with the broader pattern of western ideas emerging from Europe at the time. The sovereignty that was essential to the modern nation state was 'linked to the replacement of "frontiers" by "borders"' (Giddens, 1990, p. 73). This cultural and political shift is also reflected in the shift from earlier frontier exploitation of forests to a need to manage fixed extents of forest resources contained within national borders. The frontier is a place where expansion is always possible but a border has no such potential. A new way and new logic was required. It follows that there is an overlap with the emergence of the nation state and stewardship forestry – the logic of frontier expansion is replaced with the logic of living within fixed boundaries. The latter requires a management of a contained (and therefore limited) resource – the logic of stewardship forestry and its pursuit of sustainability.

German foresters were particularly influential in spreading stewardship forestry ideas throughout the world during the nineteenth century (Simmons, 2008, pp. 125–126). The first Inspector General for forests in British India was a German, Sir Dietrich Brandis (Advani cited in Saldanha, 1996). Along with another senior forester of German descent, Berthold Ribbentrop, Brandis was influential in advancing American forestry. He provided collegial support, advice and mentorship to the three key foresters who would oversee the establishment of the vast US public forest estate and its accompanying bureaucracy: Franklin Houghes, Charles Sargent and Gifford Pinchot (Barton, 2002). This process was replicated in other parts of the world as the expanding European influence of modernisation and industrialisation took place.

This saw the stewardship forestry model promoted and adopted throughout the European colonial sphere of influence (in the late eighteenth and early nineteenth century). The overlap of this form of forestry and colonialism led to dispossession of indigenous peoples by colonial powers – forestry was brought to the service of territorial control (Bryant, 2002; Saldanha, 1996), even though this was often in the name of sustainability, preservation, and water and climate benefits. In many places traditional communal control of forests shifted to state control. This has been called the 'Tragedies of the Invasions' (Brightman, 1987, pp. 133–134). Ghosh (2007, p. 63) laments that from 'the very ancient days Indian people lived with the forests and their interaction with forests did not do any harm to the environment', but this connection was broken with colonisation. Post-colonial analyses such as these illuminate the extent to which colonial and capitalist power relations co-opted stewardship forestry institutions to their agenda.

Also vital to the story of wood production in this period was the shift to fossil fuels as an alternate source of energy to wood (Simmons, 2008) and at the same time much more efficient commercial shipping allowing certain wood products (notoriously bulky) to be traded over larger distances. Fossil fuel use in the United Kingdom rose from its first utilisation before 1600 to one third of energy output by the end of the century (Williams, 2002, pp. 184–186). From about 1670 this boom in fossil fuel use lessened some key concerns about wood shortage that had featured in the early decades of the seventeenth century. It is also worth noting that city expansion in the period before fossil fuel use was dependent on (and limited by) availability of wood for fuel. Simmons (2008) notes that transporting wood for fuel was a costly undertaking and contributed to high wood fuel prices in early cities.

These high prices, in turn, would have provided incentive for the shift to fossil fuel sources. In this sense, the emergence of fossil fuel use in the industrial period could be seen to have relieved pressure on forests from an even higher demand for wood. Also of note, the German economist Werner Sombart argued that the drop in wood supply arising from the loss of European forest and the subsequent wave of coal-based innovation was the basis for the development of capitalism in the nineteenth century (Sombart 1913, cited in Reinert and Reinert, 2006, p. 72). Industrialisation and the concurrent rise of capitalism as an organising force for expanding and developing society have had significant implications for wood production and forest use beyond Europe.

From its beginnings in Britain, industrialisation spread to Western Europe, the United States, Russia and Japan, and thence to most of the world. It spurred a massive increase in wealth, population growth and resource consumption (Darwin, 2008). With improvements in shipping and processing, the demand for wood was no longer confined to the environs of urban centres or industrial heartlands, but was internationalised. The industrial 'core' commenced drawing wood from the world's forests to meet its needs. As manufacturing industrialised, the less developed world's share of manufacturing fell from 73 per cent in 1750 to just 6.5 per cent in 1953 (Williams, 2002, p. 246). Thus, manufacturing outputs of newly industrialised countries replaced those of less developed countries which, in turn, had to pay for manufactured goods with the export of primary produce. Among other things, this meant wood from forests and agricultural produce from land cleared of forests. Industrialisation was a key factor in the world system of forest use – this increase in pressure on forest resources also reinforced the spread of the sustainable yield ideas of stewardship forestry which often, as they did in Europe, played 'catch-up' in a period after the most readily accessible forest resources had been plundered. Vogt *et al.* (2007) summarised this process of shifting wood/forest use as follows. They identify three main uses of wood: fuel, paper products and industrial wood products, with two key trends noted in relation to this. First, nations use a significant portion of their wood as fuel until they become industrialised and, second, as countries become more industrialised they shift fuel sourcing from wood to fossil fuels. Vogt *et al.* also suggest that with industrialisation and energy from fossil fuels, more wood is freed up for consumption in timber and paper products. Put another way, when societies industrialise, wood use shifts from being primarily a source of fuel to being a source of wood for paper and timber.

Raw wood is a basic input to wood product manufacture and there are strong economic incentives to reduce the costs involved in raw wood supply, inducing innovation in wood processing. This process is having success in reducing demand for wood from forest ecosystems. Ajani (2011. p. 53) refers to the 'uncoupling of wood from finished wood products through wood saving' in her review of the divergence of stagnant world log production versus the growing consumption of wood products. Innovation in industry is focused on saving on the limiting factors in production (especially those with the highest costs), which includes raw logs and labour (Ajani, 2011, p. 53). As an example, for the period 1980 to 2007 global paper consumption grew by 3.2 per cent per annum while the use of wood pulp to meet this demand only grew by 1.4 per cent per annum. The shortfall was made up by a 5.2 per cent per annum growth in recycled paper and a 3.0 per cent per annum growth in non-wood pulp sources (Ajani, 2011, p. 56). Hyde (2012) notes that global demand for pulp logs grew by just 10 per cent from 1980 to 2000, while paper and paperboard production grew by 90 per cent in the same time period. The volume of logs harvested has not needed to keep pace with increases in the consumption of wood products. This is born out in international data showing per capita wood consumption declining consistently over previous decades despite increases in per capita economic wealth and consumption (Sewall, 2003). Conversely, changing technologies (along with growing populations) are likely to have contributed to the increase in supply of wood from forest exploitation in developing tropical countries in the later part of the twentieth century (Shearman *et al.*, 2012). Overall, then, technology acts to change demand and supply of wood in contradictory and complex ways.

Another significant feature of industrialisation's impact on wood use has been the rise of cities. Simmons (2008, p. 11) suggests that about 85 per cent of the infrastructure of cities now consists of mined products. This points to a key driver in controlling wood demand – the dramatic urbanisation of the world's population. There were cataclysmic fires in emerging urban areas built extensively of wood, such as those in London, Copenhagen and Tokyo in the fifteenth and sixteenth centuries. These helped shift city building to more durable and fire-resistant materials – that is, away from overreliance on wood. Some evidence suggests that cities are generally more resource efficient per capita than rural areas (Dodman, 2009; Newman, 2006). If so, it may be that the trend to increasing urban populations will continue to act to reduce per capita wood use.[5]

Overall, industrialisation has had a number of impacts on wood production and forests. It has seen the rate of deforestation and forest degradation from wood extraction grow, initially around the new centres of industrialisation and then across the globe as the global economic system accessed the periphery for wood. Industrialisation and related new technologies in health and food production induced massive population growth as well as dramatic increases in wealth that have raised demand for wood. Though this had extensive negative impacts on the world's forests, two aspects of industrialisation have conversely worked to potentially lessen demand for wood. First, fossil fuels as an energy source substituted large portions of earlier demand for wood. Second, cities (mandated by industrialisation and made possible by extensive use of fossil fuels) have led to forms of living that require a smaller per capita wood demand than more dispersed rural living. Finally, shifts in population, land use, and levels and distribution of wealth have allowed forests in some parts of the world, around early centres of industrialisation, to regenerate. In this period of industrialisation a major shift began to occur in wood production that would influence forests and wood production in the coming centuries across the globe – the shift from opportunistic exploitation of standing natural forest with little or no management to stewardship forestry and the emergence of the cultivation of trees for wood (Vogt *et al.*, 2007, pp. 26–27).

As already noted, the rise of scientific forestry dovetailed with the expansion of western capitalism. 'At a deeper epistemic level the language of scientific forestry worked to justify the shift towards commercial working' (Guha, 2000, p. 60). The first approach in European stewardship forestry was focused on tree plantings of species optimised for sustainable wood production. It is the logical response to the problem that drove the establishment of the institutions of forestry in the first place: how to manage forests so that wood supply is guaranteed in the long term? The Faustmann model exemplified this thinking. Developed by the German forester Martin Faustmann (1849), it determines the financially optimal rotation age by taking into account opportunity costs and discount rates (Brazee, 2001). This approach was utilitarian and commercially rational. However, this approach came to be increasingly challenged in the twentieth century as non-wood values in forests became more important. Increasingly, the Faustmann model has been challenged. For example, it is likely to determine optimal rotation ages that are less than that which deliver the optimal biological yield (Raunikar and Buongiorno, 2007). Such straight measures of wood generated are not always effective measures of the goods that forests can contribute to welfare. The changing valuations of non-timber goods have challenged the management of sustainable yield. Indeed, sustainability for forests is difficult to determine, given such changes in social values (Newman and Wagner, 2012). Newman and Wagner note how in the US public forest harvesting grew, then crashed, as changing values impacted on harvests through a range of non-market processes.

So, the initial utilitarian approach of stewardship forestry to sustainable wood production in forests set the scene for its eventual struggle with emerging demands for non-wood values from forests. In particular, this utilitarian and commercially focused approach to wood production was then imposed on natural forests that would come to be increasingly valued for a broad range of other, often conflicting, values. It is to meet this challenge that much effort has been directed in the subsequent evolution of the institutions of stewardship forestry.

Stewardship forestry – 'balance' and multiple use

The institutions of forestry evolved in response to concern about wood shortage and the need to manage forests accordingly. However, another social phenomenon, the 'ecological impulse', which rose in response to western industrialisation (Hay, 2002) was also important in shifting perceptions of forests. Certainly, by the second half of the twentieth century there was a growing recognition that the world's forests were not an inexhaustible source of wood. While the realisation was not new, an emerging global perspective could see that the frontiers had been explored, the extent of the resource was known, and its limits fixed. During this time the importance of the world's forests in providing ecological services, many with global scope, became clearer. It was in this context that stewardship forestry began to grapple with incorporating ideas of multiple values and ecological forestry into its practice. This led to variations on traditional wood focused forestry such as *multiple use forestry* and *sustainable forest management*. This was increasingly backed up with new regulatory frameworks. For example, this new turn in the stewardship forestry approach found explicit expression in the Multiple Use Sustained Yield Act of 1960 in the United States (Lane and McDonald, 2002). The management of wood production forests was legally defined as being for multiple purposes. Other legislation in other jurisdictions also reinforced this need to consider a suite of non-wood values in forest management. This approach of 'balanced' forestry was widely adopted by the formal institutions of forestry worldwide. A challenge, however, has been that the point of balance has been constantly shifting as new science, changing economic drivers and new social values continue to alter the balance equation.

This approach of multiple use and multiple values management is further exemplified by the rise of third party certification in the last decades of the twentieth century, representing a new form of governance (Cadman, 2011). It emerged, in part, due to the failures of state and international governance to regulate the often distant impacts of global trading regimes that spanned sovereign states to the satisfaction of civil society, notably environmental and social justice movements (Bartley, 2003). Certification governance is an extension of this newer stewardship forestry phase. That is, it develops to further multiple value forest management where stewardship forestry approaches stagnate when co-opted by state and capitalist interests. The 'certification wars' that have ensued are a tussle between the pull of the highest aspirations for forest care from NGOs in particular and the timber industry, keen to maintain the status quo, but also in need of a social licence (Humphreys, 2006).

It is useful to recognise that, as well as this solution of multiple-use within stewardship forestry, there were advocates for a shift to wood cultivation outside existing forests (especially plantations) in order to optimise other values in forest ecosystems. Edward Swain, who was significant in the early years of stewardship forestry in Australia, led state forest agencies; first in the state of Queensland from 1918 to 1932 and then in New South Wales from 1935 to 1942. He was a strong advocate for a model of forest management that involved the two-pronged approach of 'the creation of rapid-growing, intensively managed plantations and the preservation of natural forests for the spiritual and psychological needs of modern society' (Bennett, 2010, p. 325). This approach of planting highly productive plantations has been widely adopted in Australian forest management policy. Successive waves of plantation establishment[6] have dovetailed with a growing forest ecosystem conservation reserve estate. Despite this, it has not always been clear that the establishment of these plantations in Australia was for the rationale Swain advocated – other drivers included simply growing total national wood production and concern that natural forest ecosystems were not capable of meeting future demand.[7]

Also in this context of increasing demand for non-wood values, there was an emerging recognition of the failure of stewardship forestry to deliver its basic goal of sustainable wood production from extant forest ecosystems. In a number of places where the stewardship approach was used for forest management (such as Australia, the Pacific North West of the US, Malaysia and Canada) much wood production was taken from first cuts in ecologically mature forests with large standing volumes that were known to be unsustainable rates of extraction (Hyde, 2012; Putz *et al.*, 2012). The ability of this model to deliver effective long term multi-value forest sustainability has been challenged (Lindenmayer and Laurance, 2012; Putz *et al.*, 2012; Richardson and Peres, 2016). In response, in recent decades there have been numerous attempts to further develop the multiple-use forestry model and to find further ways to harvest wood and reduce its impacts on non-wood values (see, for example, the example of retention forestry in Gustafsson *et al.*, 2012). These generally involve some variation on limiting the intensity of logging operations – such as increasing buffers of unlogged land around sensitive landscape features such as streams or key habitats, or maintaining certain forest structural elements, such as mature trees with hollows, old woody debris or a mix of age classes. However, these measures act to increase the costs of logging through economically inefficient restrictions on harvest practice and more dispersed harvest ranges. Sutton (1999, p. 102), in noting this dilemma, predicted that the tension would be resolved through increased use of cultivated wood. He also suggested that this would allow the remaining natural forests to be set aside for non-wood uses such as wilderness and biodiversity. For him, this made sense given the high cost of natural forest management and the conflicting demands on 'multiple-use' forest managers, 'some [demands] of which are almost incompatible' (Sutton, 1999, p. 102). More bluntly, Bael and Sedjo (2006, p. 5) proclaim '[i]nnovations in tree growing have rendered reliance on existing natural forests increasingly unnecessary.'

Post-forestry – wood cultivation

Wood production is now making significant shifts from forest ecosystem extraction to the cultivation of trees outside of forest ecosystems. This has been occurring rapidly in the last several decades. Overall, a tipping point has been passed 'with the ascendancy of planted over natural forests for supplying' wood, and even some non-wood values (Evans, 2009, p. 1). Yet, even though this seems a relatively recent phenomenon there are historical references to the deliberate planting of trees (Powers, 1999).

Reaching back in time are numerous examples of the deliberate planting of trees for wood production. *Acacia nilotica* was used in ancient Egypt for boat building (Simmons, 2008, p. 85), whilst Evans (2009) includes several examples of wood cultivation from ancient times – in the Mediterranean, the Middle East and Asia. Many of these are for trees that had uses other than wood, such as olives and frankincense. Chinese efforts, dating back to the first millennium BC, included programs of reforestation. Java has a long history of teak cultivation, with the species generally considered to have been naturalised there through intentional planting for wood – its presence being well established by the time the Dutch arrived in the seventeenth century (Pandey and Brown, 2000). From the Middle Ages on, there are examples of European jurisdictions undertaking the planting of trees for wood (in Germany, England and France), initially with a focus on local species (Evans, 2009) and, as noted earlier, this was often in conjunction with early expressions of modern stewardship forestry. The nineteenth century saw the beginnings of more extensive plantation establishment in Europe. The focus was shifting to species that were faster growing. There was generally a favouring of conifers over broadleaf species and the increasing use of exotics, with plantings of North American and Asian species such as Douglas Fir and Japanese Larch (Evans, 2009; Simmons, 2008).

It is useful to recognise the overlap of stewardship forestry and wood cultivation as the two approaches developed in Europe. Only later in other parts of the world has wood cultivation progressed as a more proactive response to the declining productivity of overcut natural forest estates rather than in response to complete deforestation. Additionally, in the tropics there are examples of wood cultivation arising directly from swidden approaches (Harwood, 1996). As wood becomes scarcer and its price increases, tree cultivation as part of a mixed agroforestry system can create long-term returns to complement shorter rotation crops. This description suggests a plausible explanation for the evolution of teak cultivation in Java.

In their review of global plantation expansion in the early 1980s Sedjo and Lyon (1983) commented on the dramatic growth. They also noted that global wood supply was still dominated by extraction from the endowment of the slow-growing but extensive boreal and temperate forests of Russia, Canada, Scandinavia and the United States. At the time 70 per cent of the world's wood production was coming from these temperate regions of the northern hemisphere. The tropics and southern countries played a minor role in world production and international exports of industrial wood. Sedjo and Lyon, however, proposed that these parts of the world would have a comparative advantage in fast growing industrial plantation wood in the not too distant future. This would see wood production shift to the 'South', as indeed it has with the growth of wood production from plantations in countries such as New Zealand, South Africa, several South East Asian countries and South America, including Chile and Brazil. It should also be noted that in some cases this shift has not been between countries. In some larger countries, the global shift of wood production they describe has occurred within the country (and not necessarily as a general shift towards the tropics). In Brazil, wood production has moved from the equatorial Amazonian forests to the southern Atlantic plantation region, while in the US wood production has shifted from the Pacific North West to the South East.

This process of change is being driven by well-known technology and economic drivers. There are similar efficiency drivers in both agriculture and forestry that operate to shift production systems to intensification (Victor and Ausubel, 2000; Vincent and Binkley, 1993). Binkley (2003) noted that exploiting standing volumes in natural forests causes increased scarcity, which raises the costs of wood. He then outlined three countervailing factors that serve to mitigate this. First, as the cost of extracting wood from one area's forests grows as a result of accessing increasingly marginal forests, wood production can shift to other regions of the world. Second, increasing prices for wood make the cultivation of trees more economical. Binkley notes types of stewardship activity, such as wildfire suppression and methods of forcing suitable natural regeneration, as a first step in the application of the stewardship approach with a tendency for them to evolve into 'sophisticated agronomic practices' (Binkley, 2003, p. 2). The third factor serving to reduce costs are technological advances, increasing yields from cultivated trees and wood processing. In a similar analysis Sedjo and Botkin (1997) suggest three drivers for the transition to intensive forest management (of which plantations were a key component): rising costs associated with increasing difficulty in accessing frontier forests; technology upping the yields of planted forests; and increasing social pressure to protect old growth forest.

There is a rapidly growing global estate of cultivated wood. Brown (2000) estimated that in 1995 there were 123.7 million hectares of wood plantation supplying 22.2 per cent of the world's industrial roundwood and 4.4 per cent of its wood fuel production (see also Del Lungo *et al.*, 2006). The FAO estimated that in 2012 a conservative 33 per cent of the world's industrial roundwood was coming from plantations and 47 per cent from its planted forests, including 'semi-natural' forests (Jürgensen *et al.*, 2014). In the later centuries of the second millennium, global plantation establishment rose from very low levels in Europe and Asia. The twentieth century saw it increase, first in countries such as South Africa, the United States, New Zealand and Australia. In the last decades of the twentieth century there has been dramatic growth in developing countries (Evans, 2009, pp. 5–22; Varmola *et al.*, 2005). Fifty years ago little of the global industrial roundwood supply came from plantations. But in the second half of the twentieth century the shift to cultivated trees as a wood source has gathered pace (Evans, 2009). During the first decade of the twenty-first century the world's plantation estate grew at 2.8 million hectares a year. In 2012, 3 per cent of the world's forests were production plantations, but they produced one-third of the world's industrial roundwood (see Figure 2.2). Predictions for plantation wood suggest it will grow to more than 40 per cent of the world's industrial roundwood output by 2030, and as high as 75 per cent by 2050 (Kirilenko and Sedjo, 2007).

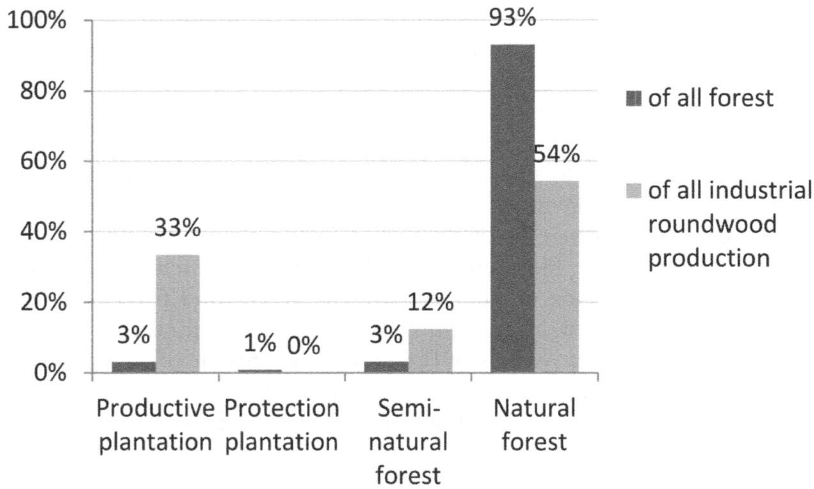

Figure 2.2 World forest extent and industrial roundwood production in 2012.[8]

One of the significant advantages that plantations offer over natural forests is their ability to grow relatively large volumes of logs in much smaller land areas with a consistent size, thereby allowing for more efficient transport and processing (reducing transport and processing costs). Technology is key to these advantages. Technology has made tree harvesting easier, with chainsaws and heavy machinery being obvious examples. And, also in the area of wood processing there has been ongoing improvement.

In the latter half of the twentieth century commercial wood processing operations became more integrated. Parts of logs that were once wasted were increasingly used in secondary products such as pulp, energy and reconstituted wood products. The Vice President of Weyerhaeuser, speaking to the Forest Products Society in 1975, observed that in 1948 only 20 per cent of their logs in Western Oregon were turned into usable product. By 1963 this was 61 per cent and by 1973 it was 79 per cent (Meil *et al.*, 2007, p. 84). In the 30 years from 1975 to 2005 significant efficiencies continued to be made in the North American wood products industry. Meil *et al.* (2007) calculated that 15,000 hectares of annual forest harvest were avoided as a result of these efficiencies. And it was not just better log utilisation – an ability to utilise smaller and poorer quality logs so that harvest activity can be more contained with lower harvesting costs. These technological changes act to reduce general demand for wood. In addition, they also act to shift demand to wood cultivation through the draw of efficient mechanised harvesting of uniform stands of trees in wood plantations (Lucier *et al.*, 2001). Hyde (2012) makes the observation that as wood product manufacturing has become much more capital intensive, the incentive for those capital owners to have uninterrupted wood supplies has increased. Supplying streams of consistent quality wood is a distinct comparative advantage for industrial plantations over natural forests. This in turn 'locks in' a mutually supportive pattern of co-location of processing and wood cultivation investments.

Pulp and paper production has particularly been pushed to sourcing its wood from cultivated sources through the economic and technological advantages of cultivated wood. The shift also sees a geographic expression as pulp production moves to new large mills in Asia and Latin America in a drive to reduce production costs (Lebedys and Li, 2014). Highly productive plantations in these tropical and subtropical locations can produce greater volumes of wood[9] from smaller areas. This in turn reduces land needs and haulage costs across the reduced catchment area needed to supply a mill.

Generally, wood volumes produced per unit land area are much higher from plantations than from natural forests. A summary of published yield tables in Paquette and Messier (2009) shows a range of mean annual increment[10] of 5–40 m^3/ha/year for intensive monoculture plantations compared to 1–3 m^3/ha/year for intensively managed natural forests, and less than 1 m^3/ha/year for certified natural forests. Work on Brazilian plantations indicates an average mean annual increment for its eucalypt plantations of 40 m^3/ha/year – up from an average of over 30 m^3/ha/year in 2000 (Gonçalves *et al.*, 2013). Häggblom (cited in Nilsson and Bull, 2005, p. 10) estimated, '88 million hectares of successful fast growing plantations of Brazilian type (mean annual increment 20–25 m^3/ha/year) would take care of the total supply for the global consumption of industrial hard wood in 2015'. In the period 2006–2012 Brazil established around a half a million new hectares of eucalypt plantation a year with an average mean annual increment in the order of 40 m^3/ha/year (Gonçalves *et al.*, 2013). Even away from the tropics, in the cold regions of north Europe fast growing plantation crops of willow and poplar species can produce 15 m^3/ha/year compared to 1.5 m^3/ha/year in the natural forests of the region (Mola-Yudego *et al.*, 2017).

An increasingly important dimension of global wood supply is wood from trees outside forests. Agroforestry planting shows signs of rapid growth. For example, planting began for fuelwood on a large scale in Africa and Asia in the 1970s (ABARE 1999, citing Panday 1995; Hyde *et al.*, 1996). The FAO calculated that this mix of cultivated trees outside forests – in agroforestry landscapes, alongside roads and canals, across farms and around villages, towns and cities – is now likely to supply most or all of the demand for fuelwood in developing parts of the world such as Africa (FAO, 2001, 2010). Another analysis estimated that two thirds of global fuelwood consumption was supplied from trees outside forests (Smeets and Faaij, 2007).

In addition, in recent decades increasing use has been made of the wood from trees grown primarily for other crops, such as rubber trees, coconut and palm oil trees (Durst *et al.*, 2004). Rising prices have generated markets for the wood of trees that would once have been burnt. Combined with the low-cost labour of the developing countries in which these crops are found and new technologies for wood processing that can utilise smaller pieces of wood (or even its constituent fibres), these provide considerable new volumes of wood resource. For example, Durst *et al.* (2004) reported that Malaysia's $2 billion furniture industry had shifted to using rubber wood as its preferred timber, supplying 80 per cent of its raw material.

As well as sourcing wood from outside forests and making better use of trees grown for other crops, alternative fibre sources are being considered. For example, bamboo is increasingly used to manufacture 'high-tech' substitutes for wood products as well as being a key landscape restoration species (Rebelo and Buckingham, 2015). It enjoys many of the advantages wood enjoys over metal, plastic and concrete materials – that is, it is low carbon, 'natural' renewable image – while often being more productive and stronger. 'It is quickly changing its image from the "poor man's tree" to a high-tech, industrial raw material and substitute for wood', observed Lobovikov *et al.* (2007) – noting also the spread of productive bamboo from its Asian origins to Latin America and Africa.

As well as expanding the supply of wood (and its constitutive fibres) through new sources such as bamboo, plantation palms and rubber trees, the productivity of cultivated trees is growing. Much as the productivity of agricultural food crops has shown a long-term propensity towards increasing yields (Tilman, 1999), so too has the productivity of cultivated trees (Nambiar, 1999). In combination, these processes further underscore the persistent direction of change towards wood cultivation to supply wood and fibre demands.

But it is not only new sources of wood and more efficient processing that is pushing increases in wood cultivation. Political forces are also tending to act to support this expansion. In the latter twentieth century there have been repeated government interventions to support early-stage development of tree cultivation.[11] This is often driven by political imperatives such as national independence and self-reliance (in the case of Australia for example, see Ajani, 2007; Gerrand *et al.*, 2003), rather than expressing any policy preference for a change in wood production methods. But as noted above there have also been forest managers who have argued for this as a way of delivering better natural forest ecosystem management. But whatever the political drivers (and of course it can be a mix of these) the effect has been significant government investment in the early stages of wood plantation.

Significant examples of this include Brazil, Indonesia, China, Vietnam and Australia. Brazil had tax incentives operating until 1986 for plantation generation that, in turn, established an industry that is now growing at even higher rates (Gonçalves *et al.*, 2013). The Indonesian government has had a number of policies in place to grow its plantations (Obidzinski and Chaudhury, 2009). The Chinese government worked to a target of 13.3 million hectares of fast-growing high-yield plantations to be established between 2001 and 2015, with 45 per cent to be pulp logs (Barr and Cossalter, 2004, p. 269). The State Forest Administration subsidises financing to encourage fast establishment.

Ajani (2011, p. 59) reported that by 2008 the country was on target to meet this goal, concerns about processing capacity notwithstanding. Vietnam's plantation estate grew at 5 per cent per year during the period 1990 to 2005 (FAO data cited in Sandewall *et al.*, 2010, p. 567). This rapidly growing estate has moved into production over the last ten years. Data cited in Phuc (2011) shows Vietnamese hardwood chip exports growing from about 500,000 m^3 a year in 2003 to nearly 4,000,000 m^3 in 2010. Sandewall *et al.* (2010, p. 567) note that the reforestation trend in Vietnam and China in particular has been significantly driven by farmers. The shift of forest use in Vietnam to plantations in some ways reflects the Australian experience where areas initially logged and then subsequently cleared for agriculture (dairy and cropping in Australia, shifting agriculture in Vietnam) have subsequently been replaced by state-promoted plantation schemes.

But the expansion of wood cultivation is not without problems. Extensive plantation establishment initiated under government-supported programs and, often, large corporate investments, has also fuelled considerable political controversy and has sometimes led to perverse outcomes. Strong criticism has been levelled at plantations on the basis of their impacts on social and ecological values (Bowyer, 2001; Casson, 2004). In Indonesia, significant areas of the land allocated for plantations since the 1990s have been cleared but not replanted. The companies involved have been using the subsidies and land allocations as a source of cheap wood in clearing the land, but then not replanting. This had the perverse outcome of a reforestation project leading to net deforestation (Obidzinski and Chaudhury, 2009, p. 83). A similar outcome occurred in the Australian state of Tasmania under a federal government tax incentive, the Managed Investment Scheme that fuelled particularly high rates of clearing of natural forest for conversion to wood plantation (Forest Practices Authority, 2014). A study of plantations established across 1.4 million hectares of Sumatra found that four of the five plantation concessions assessed failed to deliver economic benefits that were more than the economic costs (Maturana, 2005). This was primarily because the Indonesian government did not pass on the full cost of the forest lands handed over (largely areas of forest degraded through past logging). In addition, there are often social impacts implicated with the dramatic landscape and socio-economic changes associated with large scale plantation establishment (Gerber, 2011).

There has also been considerable criticism of plantations and their environmental impacts (Sawyer, 1993; Stephens and Wagner, 2007). Among these are: impacts on soil condition; water quantity and quality; chemical use; the spread of exotic species; creation of ecological 'deserts'; and the loss of natural forest through conversion. However, in many cases the criticisms point to problems of distinctive locations and management processes. There can often be considerable scope for changes and improvements in management practices to manage and reduce these impacts. Certainly, the progression of literature in relation to environmental impacts of plantations follows a pattern of swiftly expanding plantation estate, emergent problems, and then an active research and adaptive management response. This is noticeably so in relation to soil nutrition. Soil nutrition is potentially depleted when fast growing tree crops are regularly harvested and the biomass removed (Berthrong *et al.*, 2009). Yet, tree plantations may have the opposite effect, contributing to 'long-term improvement in soil quality and site productivity, especially on degraded soils ...' (Fox, 2000, p. 187), and even in examples of declining soil fertility silvicultural practices are quickly refined such that this process is reversed (O'Hehir and Nambiar, 2010). At the other end of the spectrum, issues of competing land use, conversion and the deforestation of natural forests have proved much more intractable.

Where plantation establishment involves the clearing of natural forests there is a significant biodiversity and ecosystem service cost, either directly or through forcing food production to displace natural forests (for example Bremer and Farley, 2010; Brockerhoff *et al.*, 2008; Gibbs *et al.*, 2010; Pawson *et al.*, 2013). However, Sedjo and Botkin (1997) argue that this critique fails to recognise the opportunity for vastly more productive, intensively managed forests to replace natural forests as a source of wood, allowing remaining natural forests to be better used for non-wood values such as biodiversity. Nonetheless, there are examples (such as in Indonesia and Australia) which clearly illustrate that public policy can have the effect of plantations being established at a significant cost to natural forests.

Further, in regard to biodiversity, plantations are most likely to contribute to restoration of biodiversity when established on degraded lands, rather than when replacing natural ecosystems (Bremer and Farley, 2010). The environmental outcomes of plantations vary widely depending on the type of plantation and previous land uses. For example, plantations utilising indigenous species have a more beneficial effect on biodiversity than those that do not (Bremer and Farley, 2010; Lugo, 1997). Bremer and Farley also note the importance of plantations to biodiversity conservation and recuperation at a landscape level. They found the impact on the biodiversity outcomes for plantations were negative where the plantation was established through conversion of natural forests, grasslands and shrublands. In contrast, in some cases the conversion of secondary and degraded forest and exotic pasture to plantations using indigenous species could be positive (see also Brockerhoff *et al.*, 2013).

There are also analyses that stress the considerable opportunities for utilising existing and newly cultivated wood crops to deliver a wide range of ecosystem services, while supplying roundwood (see Bauhus *et al.*, 2010, for a detailed treatise). Where these plantations are established on previously cleared land and have good management practices they have become positive examples of the possibilities of plantations. This is important because while wood cultivation production has been growing and subject to the above criticisms and demands for more positive environmental and social outcomes, such criticisms and demands reflect the rise of much higher expectations for ecosystem service delivery across the landscape.

Perhaps one of the fastest changing environmental considerations for wood and forest social-ecological systems is climate change. Cultivated trees are likely to offer greater flexibility for wood production in a world of fast changing climatic and biospheric conditions. Sohngen *et al.* (2001) predicted that climate change would increase global wood production, lower prices and increase welfare from these markets. They also found it would further the trend of wood production, shifting from primary natural forests in high latitudes to plantations closer to the equator, as predicted by Sedjo and Lyon (1983). The short rotations and ability to shift production to more suitable climates and regions give significant advantages over natural forests, which are relatively fixed as to where they exist. In addition, increased demands on natural forests to supply non-wood ecosystem services in response to climate change could further push extensive natural forest ecosystem management towards these values at the expense of wood production. Climate change mitigation measures are also likely to include demand for an increased forest estate. Where this is met by the planting of new trees on previously cleared land it is likely that part of this will be through species and plantings that are suitable for timber, further increasing future supplies of cultivated wood.

Post-forestry – forest ecosystem management

Forests have long been used for values other than wood, in particular hunting and as a source of other natural materials, such as honey and gathered foods and, indeed, continue to provide food security to significant, if declining, human populations (Parrotta *et al.*, 2015). However, more generally, as agricultural societies advanced and industrialised there has emerged a dichotomous view of forests as places separate from the cleared agricultural land from which sustenance was derived. In part, this has its origins in religion-sanctioned suspicion of forests, as displayed in the ordering by a Benedictine Abbot to bless the wood cutters because '... a wild spot ... is, as it were, in a state of original sin' (quoted in Leclercq, 1911, p. 136). But as this dichotomous view of forests developed and the drive to clear forests for agricultural land and harvest wood increased pressure on forests, a counter current in conceptions of forests has also emerged. This sees forests as reservoirs of positive virtue and value: as places such as hunting grounds for the well to do; a symbol of wealth both of the individual and the state in general; and as links to the Garden of Eden and purity. This fraught relationship with forests, in the west in particular, is described in some detail by Schama (1995). He notes a schizophrenic lurching between longings for a wooded arcadia and its sylvan inhabitants and fear of the dense forests north of the Alps and their resident barbarians.

Williams (2002, p. 166) outlines the growth of new sentiments towards forests as Europe rose to global pre-eminence. These sentiments started with utilitarian exploitation. From this, a second stage, stewardship, emerged in the face of recognition of limits to the forests and their importance. Then came a third stage – a cherishing of beauty in forests. This later stage was likely to include a growing cohort seeking outdoor recreation opportunities away from the cities they increasingly inhabited. This group of people had an entirely different set of expectations of forests, and these expectations grew in the society as affluence and recreational time spread from the upper classes through to the middle and working classes (Simmons, 2008, p. 127).

As part of these changes in how people have valued forests there has been growing recognition of the role of other services supplied by forests, particularly biodiversity, water and carbon services. These, in turn, have become part of the political-economic dynamic shaping the transition of wood production away from natural forests to cultivated trees. As Leslie (2005) notes, while demand for wood is static or only slightly increasing, demand for environmental services is burgeoning (but 'unmonetised'). This puts significant pressure on any status quo on stewardship forestry management of forest ecosystems. This is reflected in the use of forests for wood production becoming one of the most conflict-riven natural resource issues (Lindenmayer and Laurance, 2012).

Wood extraction can have significant impacts on a range of non-wood values, including ecology, climate, hydrology, recreation and culture values. The impacts are often greatest when the wood is harvested in a way that is quite different to natural forest disturbance processes and/or that reduces complexity in the resulting forest (Gustafsson *et al.*, 2012). Thompson *et al.* (2011) also note that when ecosystems such as forests are managed only for wood other ecosystem services tend to receive less attention. For these reasons there is a logic to separating wood production into wood cultivation outside forest ecosystems and optimising other non-wood values in extensive natural and semi-natural forest ecosystems. However, a counter argument says that utilising forests for wood is critical to paying for their management. As noted above, there has been considerable effort made through stewardship forestry approaches of multiple-use, sustainable forestry and ecological forestry to address the tension between the optimisation of these conflicting values. Sometimes this gets expressed as the trade-off involving wood production to raise revenue to pay for the management of forest and deliver those other values. For example, Gustafsson *et al.* (2012, p. 633) make the claim that '[m]ost ... forest owners will need to manage forests to supply ecosystem services simultaneously with the production of revenue from wood products to help pay for that management.' Here, the authors clearly distinguish wood harvest as fungible 'product' as opposed to other ecosystem services, with the clear implication that only the 'products' can pay for the management of the forests. This thinking is likely to reflect both the historical centrality of wood production in stewardship forestry logic, but it also misses the emerging markets and financial instruments being considered for payment for ecosystem services

(Kinzig *et al.*, 2011) as well as widespread management of extensive forest ecosystems in conservation land tenures that preclude wood harvest (651 million hectares globally in 2015 [FAO 2015]).

Another instance of the argument for continued wood production in natural forest ecosystems in the face of demands for protecting non-wood values is that wood products contain clear environmental advantages over substitute materials such as steel, plastic and cement, which embody considerable fossil fuel use in their production or are derived from non-sustainable mined sources. Sutton (1999) highlights the irony in increasing environmental pressure to protect forests from wood harvest for non-wood values when there is a clear environmental advantage to wood products over these other materials. However, the counterpoint can also be made that this demand for wood and its benefits can still be met from cultivated wood sources. Reducing wood production from natural forests and valuing wood for its environmental qualities are not necessarily contradictory then. It is possible to derive the environmental benefit of wood using cultivated wood sources while also deriving the environmental benefits of protecting natural forests from wood harvest.

This demand for more non-wood ecosystem services from natural forests and the resulting conflicts adds transaction costs[12] to the cost of wood production. This applies whether conducted through an adversarial political environment or through ever more intensive and demanding negotiation processes, such as certification. They all act to increase the costs of natural forest wood harvest. At the same time, if increasingly concentrated wood production through wood cultivation, with its higher yields and less contested range of ecosystem services, does not attract these same transaction costs, then wood cultivation has a comparative advantage over more extensive and lower yielding natural forested lands. Conversely, there are difficulties in assigning property rights for the many ecosystem services that forests provide, which in turn leads to free-rider problems as well as other potential transaction costs for ecosystem services other than wood (Zhang, 2001). However, on a larger scale, at least the direction of change has been in favour of wood production shifting out of forest ecosystems and into wood and fibre cultivation. This is regardless of the transaction costs of addressing conflict in natural forest wood production or establishing effective ecosystem service payment systems.

These last two post-forestry themes of this environmental history are, as noted earlier, occurring in concert. They influence one another and they arise from key contradictions in the earlier themes of stewardship forestry. As a number of authors have noted, there is a significant opportunity to address key global demands for biodiversity and other ecosystem services from natural forests, and to mitigate the perceived conflict with wood supply as the global plantation estate and wood saving technologies reduce pressure on natural forests. This creates different forms of wood production and extensive forest ecosystem management than have existed in stewardship forestry and its logical integration of wood production within extensive forest. The relative importance of natural forests as a wood source will reduce as wood cultivation grows. An institutional and geographic separation is occurring between forest use and the cultivation of wood. The following chapter will consider this change through a heuristic model of this transition, which provides a basis for further consideration of the sustainability and resilience of wood and forest social-ecological systems.

Notes

1 The term *hunter-gatherer* for societies that sustained themselves from foraging for food (non or pre agricultural) has been contested (Lee, 1992). Here I use *gatherer-hunter* in a nod to the complexity behind the terms and to acknowledge the gender biases implicit in the terms.

2 The basic Von Thunen model proposes that distance from a central market and the transport costs for goods to that central market determine the surrounding pattern of land use (Nelson, 2002). The model has been applied to an understanding of patterns of reforestation, wood production and deforestation (Ahrends *et al.*, 2010; Horvath, 1969; Hyde and Seve, 1993) as well as a basis for understanding forest transition (Angelsen, 2007; Rudel, 1998).

3 Swidden is a term used to describe a shifting agricultural development involving clearing of vegetation through fire and cutting, planting and subsequent abandoning of the fields as site productivity decreases. It is widely practiced in the uplands of Southeast Asia, but also parts of the Americas and Pacific. See Mertz *et al.* (2009) for details.

4 Based on the 1750 estimated extent of 4,822,000 hectares and the 2005 extent of natural forests from *Sustainability Indicators for Tasmanian Forests 2001–2006* (Tasmanian and Australian Governments, 2007).

5 Recent innovations in use of wood in multi-storey buildings could potentially be an emerging counter trend, although at this stage there is no indication that it is likely to grow into a significant additional source of wood demand. There is also some interest in the use of wood for biofuels and other new applications. It is not clear, though, whether these are driven by demand for the wood-sourced product or by wood producers attempting to develop new markets for a surplus of wood.

6 There was extensive expansion of exotic pine species plantations in Australia from the mid-1960s to the late 1980s and then a focus on eucalyptus species in the 1990s and 2000s (Ajani, 2007; Ferguson, 2014).

7 See the Australian case study in the chapter, *A natural turn: land use change, leakage and forest conservation* for a detailed description.

8 Forest extent data is estimated using data from Forest Resource Assessments of 2005 and 2015 including rates of change (FAO, 2006; FAO, 2015). Industrial roundwood production estimates are from Jürgensen *et al.* (2014).

9 Brown, in his review of plantations internationally, noted that, in general, tropical plantations have higher productivity than in temperate regions. (Brown 2000, p. xvi).

10 The mean annual increment (often called MAI) is a measure of forest productivity. It is measured in units of wood grown per unit of land per year, most commonly now, metres cubed per hectare per year.

11 See Howarth (2012) for a relevant discussion on the role of government in early-stage technology intervention in clean energy technology.

12 *Transaction costs* are the costs of 'defining, protecting and enforcing the property rights of goods' and are contrasted to the *transformation costs* of turning materials into higher value goods (North, 1990, p. 28).

References

ABARE (Australian Bureau of Agricultural and Resource Economics) (1999). Global outlook for plantations. *ABARE Research report*. Canberra: ABARE.

Ahrends, A., Burgess, N.D., Milledge, S.A., Bulling, M.T., Fisher, B., Smart, J.C., Clarke, G.P., Mhoro, B.E. and Lewis, S.L. (2010). Predictable waves of sequential forest degradation and biodiversity loss spreading from an African city. *Proceedings of the National Academy of Sciences*, 107, 14556–14561.

Ajani, J. (2007). *The forest wars*. Melbourne: Melbourne University Press.

Ajani, J. (2011). The global wood market, wood resource productivity and price trends: an examination with special attention to China. *Environmental Conservation*, 38, 53–63.

Angelsen, A. (2007). Forest cover change in space and time: combining the von Thunen and forest transition theories. *Policy Research Working Paper*. Washington, DC: World Bank.

Bael, D. and Sedjo, R. (2006). Toward globalization of the forest products industry: some trends. *Discussion paper* Washington, DC: Resources for the Future.

Bailey, R.C., Head, G., Jenike, M., Owen, B., Rechtman, R. and Zechenter, E. (1989). Hunting and gathering in tropical rain forest: is it possible? *American Anthropologist*, 91, 59–82.

Barr, C. and Cossalter, C. (2004). China's development of a plantation-based wood pulp industry: government policies, financial incentives, and investment trends. *International Forestry Review*, 6, 267–281.

Bartley, T. (2003). Certifying forests and factories: States, social movements, and the rise of private regulation in the apparel and forest products fields. *Politics & Society*, 31, 433–464.

Barton, G. (2002). *Empire forestry and the origins of environmentalism*. Cambridge: Cambridge University Press.

Bauhus, J., Van Der Meer, P. and Kanninen, M. (2010). *Ecosystem goods and services from plantation forests*. Oxon: Earthscan.

Belton, B. and Thilsted, S.H. (2014). Fisheries in transition: Food and nutrition security implications for the global South. *Global Food Security*, 3, 59–66.

Bennett, B. M. (2010). *Locality and empire: networks of forestry in Australia, India, and South Africa,* pp. 1843–1848. PhD thesis, University of Texas.

Berthrong, S.T., Jobbágy, E.G. and Jackson, R.B. (2009). A global meta-analysis of soil exchangeable cations, pH, carbon, and nitrogen with afforestation. *Ecological Applications*, 19, 2228–2241.

Binkley, C.S. (2003). Forestry in the long sweep of history. In: Teeter, L.D., Cashore, B.W. and and Zhang, D. (eds.) *Forest policy for private forestry: global and regional challenges*. Oxon: CABI.

Bowman, D.M., Balch, J., Artaxo, P., Bond, W.J., Cochrane, M.A., D'antonio, C.M., DeFries, R., Johnston, F.H., Keeley, J.E., Krawchuk, M.A. and Kull, C.A. (2011). The human dimension of fire regimes on Earth. *Journal of biogeography*, 38, 2223–2236.

Bowyer, J.L. (2001). Environmental implications of wood production in intensively managed plantations. *Wood and Fiber Science*, 33, 318–333.

Brazee, R.J. (2001). The Faustmann formula: fundamental to forest economics 150 years after publication. *Forest Science*, 47, 441.

Bremer, L.L. and Farley, K.A. (2010). Does plantation forestry restore biodiversity or create green deserts? A synthesis of the effects of land-use transitions on plant species richness. *Biodiversity and Conservation*, 19, 3893–3915.

Brightman, R.A. (1987). Conservation and resource depletion: the case of the boreal forest Algonquians. In: McCay, B.J. and Acheson, J.M. (eds.) *The question of the commons: the culture and ecology of communal resources*. Tucson, AZ: University of Arizona Press.

Brockerhoff, E.G., Jactel, H., Parrotta, J.A. and Ferraz, S.F.B. (2013). Role of eucalypt and other planted forests in biodiversity conservation and the provision of biodiversity-related ecosystem services. *Forest Ecology and Management*, 301, 43–50.

Brockerhoff, E.G., Jactel, H., Parrotta, J.A., Quine, C.P. and Sayer, J. (2008). Plantation forests and biodiversity: oxymoron or opportunity? *Biodiversity and Conservation*, 17, 925–951.

Brown, C. (2000). *The global outlook for future wood supply from forest plantations*. Rome: Food and Agricultural Organisation of the United Nations.

Bryant, R.L. (2002). The politics of forestry in Burma. In: Hirsch, P. and Warren, C. (eds.) *The politics of environment in Southeast Asia*. London: Routledge.

Cadman, T. (2011). *Quality and legitimacy of global governance: case lessons from forestry*. London: Palgrave Macmillan.

Casson, A. (2004). The controversy surrounding eucalypts in social forestry programs of Asia. Canberra: Economics Division, Research School of Pacific and Asian Studies, Australian National University.

Darwin, J. (2008). *After Tamerlane: the rise and fall of Global Empires, 1400–2000*. London: Penguin UK.

Del Lungo, A., Ball, J. and Carle, J. (2006). Global planted forests thematic study. Results and analysis. Rome: Food and Agricultural Organisation of the United Nations.

Dodman, D. (2009). Blaming cities for climate change? An analysis of urban greenhouse gas emissions inventories. *Environment and Urbanization*, 21, 185–201.

Durst, P.B., Killmann, W. and Brown, C. (2004). Asia's new woods. *Journal of Forestry*, 102, 46–53.

Evans, J. (2009). *Planted forests: uses, impacts and sustainability*. Rome: CABI International and FAO.

FAO (2001). *Global forest resources assessment 2000: main report*. Rome: Food and Agriculture Organization of the United Nations.

FAO (2006). *Global forest resources assessment 2005: progress towards sustainable forest management*. Rome: Food and Agriculture Organization of the United Nations.

FAO (2010). *Global forest resources assessment 2010: main report*. Rome: Food and Agriculture Organization of the United Nations.

FAO (2015). *Global forest resources assessment 2015: how are the world's forests changing?*. Rome: Food and Agriculture Organization of the United Nations.

Faustmann, M. (1849). Calculation of the value which forest land and immature stands possess for forestry. *Journal of Forest Economics*, 1, 7–44.

Ferguson, I. (2014). Australian plantations: mixed signals ahead. *International Forestry Review*, 16, 160–171.

Forest Practices Authority (2014). Forest Practices Authority Annual Report 2013–2014. Hobart, Australia: Forest Practices Authority.

Fox, T.R. (2000). Sustained productivity in intensively managed forest plantations. *Forest Ecology and Management*, 138, 187–202.

Garcin, Y., Deschamps, P., Ménot, G., De Saulieu, G., Schefuß, E., Sebag, D., Dupont, L.M., Oslisly, R., Brademann, B. and Mbusnum, K.G. (2018). Early anthropogenic impact on Western Central African rainforests 2,600 y ago. *Proceedings of the National Academy of Sciences*, 115, 3261–3266.

Gerber, J.-F. (2011). Conflicts over industrial tree plantations in the South: who, how and why? *Global Environmental Change*, 21, 165–176.

Gerrand, A., Keenan, R.J., Kanowski, P. and Stanton, R. (2003). Australian forest plantations: an overview of industry, environmental and community issues and benefits. *Australian Forestry*, 66, 1–8.

Ghosh, A.K. (2007). State versus people: the Indian experience. In: Chakrabarti, R. (ed.) *Situating environmental history*. New Delhi: Ajay Kumar Jain.

Gibbs, H., Ruesch, A., Achard, F., Clayton, M., Holmgren, P., Ramankutty, N. and Foley, J. (2010). Tropical forests were the primary sources of new agricultural land in the 1980s and 1990s. *Proceedings of the National Academy of Sciences*, 107, 16732–16737.

Giddens, A. (1990). *The consequences of modernity*. Cambridge, Polity Press.

Gonçalves, J.L.D.M., Alvares, C.A., Higa, A.R., Silva, L.D., Alfenas, A.C., Stahl, J, Ferraz, S.F.D.B., Lima, W.D.P., Brancalion, P.H.S. and Hubner, A. (2013). Integrating genetic and silvicultural strategies to minimize abiotic and biotic constraints in Brazilian eucalypt plantations. *Forest Ecology and Management*, 301, 6–27.

Guha, R. (2000). *The unquiet woods: ecological change and peasant resistance in the Himalaya*. Berkeley, CA: University of California Press.

Gustafsson, L., Baker, S.C., Bauhus, J., Beese, W.J., Brodie, A., Kouki, J., Lindenmayer, D.B., Lohmus, A., Pastur, G.M., Messier, C., Neyland, M., Palik, B., Sverdrup-Thygeson, A., Volney, W.J.A., Wayne, A. and Franklin, J.F. (2012). Retention forestry to maintain multifunctional forests: a world perspective. *Bioscience*, 62, 633–645.

Harwood, R.R. (1996). Development pathways toward sustainable systems following slash-and-burn. *Agriculture, Ecosystems & Environment*, 58, 75–86.

Hay, P.R. (2002). *Main currents in western environmental thought*. Sydney: University of New South Wales Press.

Hermann, R.K. (1976). Man and forests – a prodigal relation. *North West Foundation Forestry Series*. pp. 29–51. Corvallis, Oregon: School of Forestry, Oregon State University.

Hladik, C.M., Linares, O.F., Hladik, A., Pagezy, H. and Semple, A. (1993). Tropical forests, people, and food: an overview. In: Hladik, C.M., Linares, O.F., Hladik, A., Pagezy, H., Semple, A. and Hadley, M. (eds.) *Tropical forests, people and food: biocultural interactions and applications to development*. Paris: UNESCO.

Horvath, R.J. (1969). Von Thünen's isolated state and the area around Addis Ababa, Ethiopia. *Annals of the Association of American Geographers*, 59, 308–323.

Howarth, N. (2012). Clean energy technology and the role of non-carbon price-based policy: an evolutionary economics perspective. *European Planning Studies*, 20, 871–891.

Hughes, J.D. (2006). *What is Environmental History*? Cambridge: Polity Press.

Humphreys, D. (2006). *Logjam: Deforestation and the Crisis of Global Governance*. London: Earthscan.

Hyde, W.F. (2012). *The Global Economics of Forestry*. New York: Resources for the Future.

Hyde, W.F., Amacher, G.S. and Magrath, W. (1996). Deforestation and forest land use: Theory, evidence, and policy implications. *The World Bank Research Observer*, 11, 223–248.

Hyde, W.F. and Seve, J.E. (1993). The economic role of wood products in tropical deforestation: The severe example of Malawi. *Forest Ecology and Management*, 57, 283–300.

Iwamoto, J. (2002). The development of Japanese forestry. In: Iwai, Y. (Ed.) *Forestry and the forest industry in Japan*. Vancouver, Canada: UBC Press.

Jürgensen, C., Kollert, W. and Lebedys, A. (2014). Assessment of industrial roundwood production from planted forests. *FAO Planted Forests and Trees Working Paper*. Rome: Food and Agriculture Organization of the United Nations.

Kimmins, J. H. (2002). Future shock in forestry: where have we come from; where are we going; is there a "right way" to manage forests? Lessons from Thoreau, Leopold, Toffler, Botkin and nature. *The Forestry Chronicle*, 78, 263–271.

Kinzig, A., Perrings, C., Chapin, F., Polasky, S., Smith, V., Tilman, D. and Turner, B. (2011). Paying for ecosystem services – promise and peril. *Science*, 334, 603–604.

Kirilenko, A.P. and Sedjo, R.A. (2007). Climate change impacts on forestry. *Proceedings of the National Academy of Sciences*, 104, 19697–19702.

Lane, M.B. and Mcdonald, G. (2002). Towards a general model of forest management through time: evidence from Australia, USA, and Canada. *Land Use Policy*, 19, 193–206.

Lebedys, A. and Li, Y. (2014). Contribution of the forestry sector to national economies, 1990–2011. *Finance Working Paper*. Rome: Food and Agricultural Organisation of the United Nations.

Leclercq, J. (1911). *The love of learning and the desire for God: a study of monastic culture*. New York: Fordham University Press.

Lee, R.B. (1992). Art, science, or politics? The crisis in hunter-gatherer studies. *American Anthropologist*, 94, 31–54.

Leslie, A. (2005). What will we want from the forests. *ITTO Tropical Forest Update*, 15, 14–16.

Lindenmayer, D.B. and Laurance, W.F. (2012). A history of hubris – cautionary lessons in ecologically sustainable forest management. *Biological Conservation*, 151, 11–16.

Lobovikov, M., Paudel, S., Piazza, M., Ren, H. and Junqi, W. (2005). *World bamboo resources: a thematic study prepared in the framework of the Global Forest Resouces Assessment*. Rome: Food and Agriculture Organization of the United Nations.

Lucier, A.A., Hinchee, M. and McCullough, R.B. (2001). Biotechnology and the forest products industry. In: Strauss, S.H. and Bradshaw, T.H.D. (eds.) *Proceedings of the First International Symposium on Ecological and Societal Aspects of Transgenic Plantations*. Stevenson, WA: College of Forestry, Oregon State University.

Lugo, A.E. (1997). The apparent paradox of re-establishing species richness on degraded lands with tree monocultures. *Forest ecology and management*, 99, 9–19.

Maturana, J. (2005). *Economic Costs and Benefits of Allocating Forest Land for Industrial Tree Plantation Development in Indonesia*. Jakarta: CIFOR Working Paper No. 30.

Meil, J., Wilson, J., O'Connor, J. and Dangerfield, J. (2007). An assessment of wood product processing technology advancements between the CORRIM I and II studies. *Forest Products Journal*, 57, 83–89.

Mertz, O., Padoch, C., Fox, J., Cramb, R.A., Leisz, S.J., Lam, N.T. and Vien, T.D. (2009). Swidden change in Southeast Asia: understanding causes and consequences. *Human Ecology*, 37, 259–264.

Mola-Yudego, B., Arevalo, J., Diaz-Yanez, O., Dimitriou, I., Haapala, A., Ferraz Filho, A.C., Selkimäki, M. and Valbuena, R. (2017). Wood biomass potentials for energy in northern Europe: Forest or plantations? *Biomass and Bioenergy*, 106, 95–103.

Nambiar, E.S. (1999). Productivity and sustainability of plantation forests. *Bosque*, 20, 9–21.

Nelson, G.C. (2002). Introduction to the special issue on spatial analysis for agricultural economists. *Agricultural Economics*, 27, 197–200.

Newman, D.H. and Wagner, J.E. (2012). Putting Samuelson's economics of forestry into context: the limits of forest economics in policy debates. *Journal of Natural Resources Policy Research*, 4, 215–218.

Newman, P. (2006). The environmental impact of cities. *Environment and Urbanization*, 18, 275–295.

Nilsson, S. and Bull, G. (2005). Global wood supply analysis. *The 46th Session of the FAO Advisory Committee on Paper and Wood Products*. Vancouver, Canada.

North, D.C. (1990). *Institutions, institutional change and economic performance*. Cambridge: Cambridge University Press.

O'Hehir, J. and Nambiar, E. (2010). Productivity of three successive rotations of *P. radiata* plantations in South Australia over a century. *Forest Ecology and Management*, 259, 1857–1869.

Obidzinski, K. and Chaudhury, M. (2009). Transition to timber plantation based forestry in Indonesia: towards a feasible new policy. *International Forestry Review*, 11, 79–87.

Pandey, D. and Brown, C. (2000). Teak: a global overview. *Unasylva*, 51, 3–13.

Paquette, A. and Messier, C. (2009). The role of plantations in managing the world's forests in the Anthropocene. *Frontiers in Ecology and the Environment*, 8, 27–34.

Parrotta, J.A., Dey de Pryck, J., Darko Obiri, B., Padoch, C., Powell, B. and Sandbrook, C. (2015). *The historical, environmental and socio-economic context of forests and tree-based systems for food security and nutrition*. In: Vira, B., Mansourian, S. and Wildburger, C. (eds.). *Forests and food: addressing hunger and nutrition across sustainable landscapes*. Cambridge, UK: Open Book Publishers.

Pawson, S., Brin, A., Brockerhoff, E., Lamb, D., Payn, T., Paquette, A. and Parrotta, J. (2013). Plantation forests, climate change and biodiversity. *Biodiversity and Conservation*, 22, 1–25.

Phuc, N. (2011). An overview of Vietnam's plantations, woodchip demand, supply and outlook. *The Fifth International Pulpwood Resources and Trade Conference*. Singapore.

Piperno, D.R., McMichael, C. and Bush, M.B. (2017). Further evidence for localized, short-term anthropogenic forest alterations across pre-Columbian Amazonia. *Proceedings of the National Academy of Sciences*, 114, E4118–E4119.

Powers, R.F. On the sustainable productivity of planted forests (1999). *Planted forests: contributions to the quest for sustainable societies*. Dordrecht, The Netherlands: Springer.

Putz, F.E., Zuidema, P.A., Synnott, T., Peña-Claros, M., Pinard, M.A., Sheil, D., Vanclay, J.K., Sist, P., Gourlet-Fleury, S. and Griscom, B. (2012). Sustaining conservation values in selectively logged tropical forests: the attained and the attainable. *Conservation Letters*, 5, 296–303.

Raunikar, R. and Buongiorno, J. (2007). Forestry economics: historical background and current issues. *Handbook of operations research in natural resources*. Dordrecht, The Netherlands: Springer.

Rebelo, C. and Buckingham, K. (2015). Bamboo: the opportunities for forest and landscape restoration. *Unasylva*, 66, 91.

Reinert, H. and Reinert, E. (2006). Creative destruction in economics: Nietzsche, Sombart, Schumpeter. In: Backhaus, J. and Drechsler, W. (eds.) *Friedrich Nietzsche (1844–1900): Economy and Society*. New York: Springer.

Richardson, V.A. and Peres, C.A. (2016). Temporal decay in timber species composition and value in Amazonian logging concessions. *PloS one*, *11*, e0159035.

Ruddiman, W.F. (2012). The Anthropocene. *Annual Review of Earth and Planetary Sciences*, 41, 45–68.

Rudel, T.K. (1998). Is there a forest transition? Deforestation, reforestation, and development. *Rural Sociology*, 63, 533–552.
Saldanha, I.M. (1996). Colonialism and professionalism: a German forester in India. *Environment and History*, 2, 195–219.
Salick, J., Amend, A., Anderson, D., Hoffmeister, K., Gunn, B. and Zhendong, F. (2007). Tibetan sacred sites conserve old growth trees and cover in the eastern Himalayas. *Biodiversity and Conservation*, 16, 693.
Sandewall, M., Ohlsson, B., Sandewall, R.K. and Viet, L.S. (2010). The expansion of farm-based plantation forestry in Vietnam. *Ambio*, 39, 567–579.
Sawyer, J. (1993). *Plantations in the tropics: environmental concerns*, Gland, Switzerland: IUCN/UNEP/WWF.
Schama, S. (1995). *Landscape and memory*. London: Fontana.
Sedjo, R.A. and Botkin, D. (1997). Using forest plantations to spare natural forests. *Environment: Science and Policy for Sustainable Development*, 39, 14–30.
Sedjo, R.A. and Lyon, K.S. (1983). Long-term forest resources trade, global timber supply, and intertemporal comparative advantage. *American Journal of Agricultural Economics*, 65, 1010–1016.
Sewall, J. (2003). Long-run demand for wood – trends in world population and wood use. *Timberland Report*, 4, 1–4.
Shearman, P., Bryan, J. and Laurance, W.F. (2012). Are we approaching 'peak timber' in the tropics? *Biological Conservation*, 151, 17–21.
Simmons, I.G. (2008). *Global environmental history: 10,000 BC to AD 2000*. Edinburgh: Edinburgh University Press.
Smeets, E.M.W. and Faaij, A.P.C. (2007). Bioenergy potentials from forestry in 2050. *Climatic Change*, 81, 353–390.
Sohngen, B., Mendelsohn, R. and Sedjo, R. (2001). A global model of climate change impacts on timber markets. *Journal of Agricultural and Resource Economics*, 26, 326–343.
Sombart, W. (1913). *Krieg und kapitalismus*. Munich: Duncker & Humblot.
Stephens, S.S. and Wagner, M.R. (2007). Forest Plantations and Biodiversity: A Fresh Perspective. *Journal of Forestry*, 105, 307–313.
Sutton, W.R.J. (1999). The need for planted forests and the example of radiata pine. *New Forests*, 17, 95–109.
Tasmanian and Australian Governments (2007). Sustainability indicators for Tasmanian forests 2001–2006. Hobart, Australia: Tasmanian and Australian Governments.
Thompson, I.D., Okabe, K., Tylianakis, J.M., Kumar, P., Brockerhoff, E.G., Schellhorn, N.A., Parrotta, J.A. and Nasi, R. (2011). Forest biodiversity and the delivery of ecosystem goods and services: translating science into policy. *Bioscience*, 61, 972–981.
Tilman, D. (1999). Global environmental impacts of agricultural expansion: the need for sustainable and efficient practices. *Proceedings of the National Academy of Sciences*, 96, 5995–6000.
Varmola, M., Gautier, D., Lee, D., Montagnini, F., Saramäki, J., Mery, G., Alfaro, R., Kanninen, M. and Lobovikov, M. (2005). Diversifying functions of planted forests. Forests in the global balance – changing paradigms. *IUFRO World Series*, 17.
Victor, D.G. and Ausubel, J.H. (2000). Restoring the forests. *Foreign Affairs*, 79 (6) 127–144.
Vincent, J.R. and Binkley, C.S. (1993). Efficient multiple-use forestry may require land-use specialization. *Land Economics*, 69, 370–376.
Vogt, K., Gara, R., Honea, J., Vogt, D., Patel-Weynand, T., Roads, P., Fanzeres, A., Sigurdardottir, R., Honea, J. and Andreu, M. (2007). Historical perceptions and

uses of forests. In: Vogt, K.A., Honea, J., Vogt, D.J., Andreu, M., Edmonds, R., Sigurdardottir, R. and Patel-Weynand, T. (eds.) *Forests and society: sustainability and life cycles of forests in human landscapes*. Wallingford, UK: CAB International.

Williams, M. (2002). *Deforesting the earth: From prehistory to global crisis*. Chicago: University of Chicago Press.

Zhang, Y. (2001). Economics of transaction costs saving forestry. *Ecological Economics*, 36, 197–204.

3 The evolution of forest use and wood sourcing – a model

Wood production is undergoing a fundamental shift analogous to the evolution of agriculture emerging from earlier methods of food sourcing from natural ecosystems. At the same time, demand for non-wood ecosystem values from forests is growing. Together these processes are forcing significant change in wood and forest social-ecological systems. This process is replicated across regions and countries around the world, as well as being observable across the world system. The next section describes a heuristic model of the transition and divergence in forest use and wood sourcing evident in wood and forest social-ecological systems. This model is derived using the general idea of social-ecological system change described by resilience thinking. After setting out the theoretical basis and the model itself, this chapter considers the evidence to support the model as well as some of its limitations.

Resilience theory

As noted in the introduction, resilience thinking is a body of work describing how social-ecological systems develop states that exhibit resilience to external pressures – up to a point. A resilient state is a system condition robust to a certain level or type of fluctuation in state variables (factors or features that define the state). A particular resilient state can be known as a *basin of attraction* or simply *basin*. When a system is subject to fluctuation of state variables that is beyond the capacity of the system to maintain itself, it will shift into a new basin of attraction (a change sometimes referred to as *transformation*).

A critical feature of Holling's (1973) seminal application of the idea of resilience to ecological and environmental discourse was its challenge to the predominance of equilibrium concepts in ecological thinking (Folke *et al.*, 2002; Walker and Cooper, 2011). The concept of resilience has since been applied to a wide range of 'interdisciplinary work concerned with the interactions between people and nature' (Carpenter *et al.*, 2001, p. 765). While originally developed to apply to specific ecosystems (often lakes and wetlands) and their interconnected social systems, it has also been applied to natural resource regimes more broadly. For example, in marine ecosystems and fisheries (Young, 2010), water management (Herrfahrdt-Pähle and Pahl-Wostl, 2012), and forestry (Rist and Moen, 2013).

There are three types of response for social-ecological systems under pressure: resilience, adaptability and transformability (Walker *et al.*, 2004). The first response is the eponymous *resilience* and its four components of latitude (the elasticity, or extent to which a system can be altered but still recover to maintain its function), resistance (the ability of the system to withstand change), precariousness (how close the system is to its limits – loss of function) and 'panarchy' (the influence on the system of other enveloping systems, containing or influencing the system). The second response is *adaptability*, referring to the capacity of actors in the system to perpetuate its resilience through maintenance of its key features despite change in social-ecological system conditions. The third is *transformability*, the capacity to create a new state when the functioning of the current state becomes 'untenable'. Qualities that contribute to the resilience of social-ecological systems can be characterised as *attractors* – factors that act to stabilise or hold a system in place and act as positive feedback, and *stressors* – factors that act to destabilise a system, and potentially shift the system out of its basin of attraction and into a new state (Young, 2010).

Actors are important parts of these systems. Their dynamism can influence the resilience, adaptability and transformability of basins. They are important feedback mechanisms. They will invest in maintaining existing structures and functions that can contribute to social-ecological system resilience through a cycle of positive feedback (Pierson, 2000b). Conversely, however, actors can also resist change creating negative feedback, system fragilities or act to miss transformation opportunities during crises because of inbuilt institutional tendencies to quickly repair to a pre-crisis state (Herrfahrdt-Pähle and Pahl-Wostl, 2012). Other social attractors can include: sunk costs in infrastructure, plant, land and technologies (David, 1985); political and governance structures such as legislation, regulation and administrative arrangements or political capital (Pierson, 2000a); distributional coalitions, especially those that coalesce into forms that inhibit optimal adaptive or transformational responses in line with new system conditions (Olson, 1982); and investment of identity in social and cultural norms forged within the system, for example discourse coalitions (Hajer, 1993). These social institutions can create a degree of resilience despite the fact that changes in the state variables (for example, population, wood product demand, social values, technological change and long term economic growth) of these social-ecological systems might otherwise force a change in the system. In effect, actors and institutions can play an important part in giving the social-ecological system a form of 'lumpy' change such as that described by basins in resilience thinking.

Another change-affecting factor is that parts of a larger system (for example, a global resource use social-ecological system) can be in different basins in different places at the same time, impacting interconnected social-ecological systems. These outside, or embracing, systems form the *panarchy* of the system. For example, changes in wood production, consumption or forest use in one polity (say, increasing plantation wood cultivation) can change the social-ecological system in other polities (by, say, decreasing natural forest wood demand). The location of these social-ecological systems within different basins at national or subnational levels within the world system can also expose systems to the lessons of polities that have already faced similar challenges and developed certain responses, thus altering systems and rates of system change by an infusion of new knowledge.

These resilience concepts provide a useful framework for describing complex sets of interacting phenomena as whole systems, in a way that analysis through the silos of traditional disciplines or bounded foci on selected aspects of the larger phenomenon cannot. Understanding the crossover between natural and social systems is critical to addressing real societal problems (MacIlwain, 2014). Further, interconnected biological and social systems in forest and wood production are obvious areas for the application of ecological theory to social and economic systems – both as metaphor and analogy (Hodgson, 1993; Kant *et al.*, 2013). While the application of metaphor in scientific endeavour is open to challenge as a 'superfluous distraction in intellectual endeavour', its application can challenge 'beliefs about reality' and be a 'source of creativity' (Hodgson, 1993, pp. 18–21). Application of biological metaphors in social science has a plausible basis, given the presence of shared system characteristics. In particular, the larger systems in question have complex adaptive parts that are affected by, and in turn effect, the total system. This is fundamentally different to mechanistic systems in which the parts are unaffected by the actions or processes of the system as a whole. This chapter uses the metaphors of ecologically derived resilience thinking to define a global process of change. As applied here, resilience thinking offers a chance to tell a complex story, such that existing challenges in forest use, wood production, conservation, governance and policy can be reassessed. It also allows comparison between resilience, adaptation and transformation as alternative responses to pressure on these systems.

A model of change in wood and forest social-ecological systems

The following model (Figure 3.1) describes four distinct basins that occur in wood and forest social-ecological systems. The basins described have been located widely, historically and geographically, and can also exist in different parts of the same jurisdiction or regions at the same time. As noted earlier, evolutionary or developmental phases in wood and forest social-ecological systems have been previously explored. However, a feature of these analyses is a focus on forest use and wood production as integrated activities. Instead, this model suggests that shifts in wood production from natural forest extraction to cultivation mean wood production and forest use become geographically and institutionally distinct.

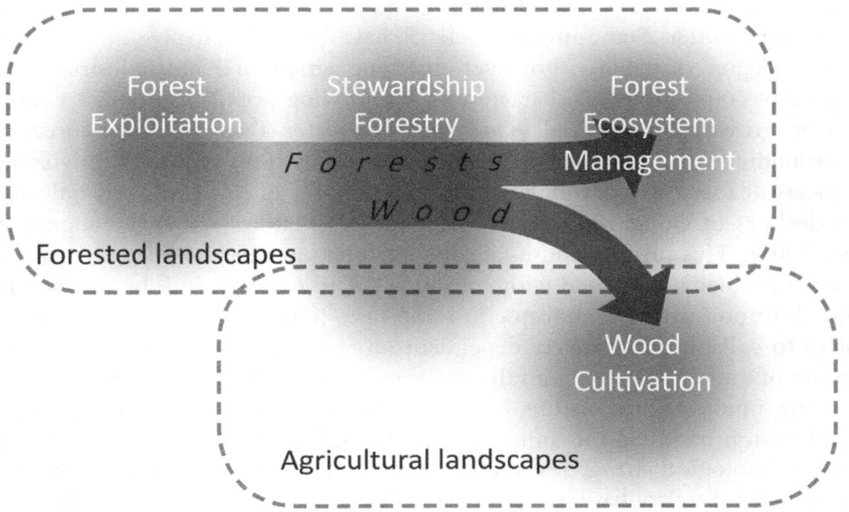

Figure 3.1 Evolving and diverging basins of attraction in wood production and forest use social-ecological systems showing a temporal sequence from left to right.

The first basin described – forest exploitation – involves the opportunistic taking of wood found in naturally regenerated forests. As noted in the previous chapter, this has occurred both as part of the conversion of forest to agricultural land and as taking of wood alone. The next basin – stewardship forestry – comes about as wood exploitation and conversion for agricultural land cause wood to become less available. Actors respond by creating institutions of *forestry* that control access to remaining forests and, combined with study of the forest and imposition of restrictions on natural forest logging, aim to achieve sustainability of wood supply. This basin sees 'wild' and extensive naturally regenerated forests on frontiers become managed and owned forests.

In time, economic and technical pressures lead to the application of increasingly agronomic models to tree growth, creating regimes of wood cultivation more akin to agriculture. This takes forms such as wood plantations, crop trees, agroforestry and trees outside forests – wood cultivation. At the same time, in response to increased societal demand for other social and ecosystem values, management of natural (and semi-natural) forests become more concerned with the delivery of non-wood values. Simultaneous with wood production moving into the wood cultivation basin, a basin of forest ecosystem management emerges that no longer has wood extraction as a central concern (or a concern at all).

There are distinctive patterns of land use associated with each of the described basins of wood and forest social-ecological systems. These patterns are not necessarily static within a basin. The forest exploitation basin in particular can involve a pattern of shifting land use as wood extraction follows a receding frontier. Some human societies have been able to live in this basin with a degree of stability, however emerging agricultural and industrial societies have typically been characterised by growing populations and a consequent growth in wood demand that is greater than the fixed area of surrounding natural forest has been able to supply, thus requiring a dynamic process of forest frontier exploitation. The impact of this pattern is reflected in declines in global forest extent and quality (Matthews, 1983; Pongratz *et al.*, 2008). This phase of unsustainable exploitation can also be accompanied by the problems associated with commons (Hardin, 1968) and the subsequent social turmoil, and even collapse in some cases (Diamond, 2005). The alternative to collapse for a society dependent on forests is the exercise of restraint in use of the resource through the creation of new social institutions.

The unidirectional nature of forest and land use change in the forest exploitation basin, and its inherent unsustainability in the face of population and technological change leads to crises that catalyse stewardship forestry. The stewardship forestry basin is characterised by its underlying logic of achieving stability of forest extent and wood production capability. Extant forests are often made the property of sovereigns, nations or private owners in a shift from frontier commons to active protection through regulated use and access and the establishment of forest management bureaucracies (Williams, 2002). The slow gradual nature of this shift of forest exploitation to stewardship forestry on a global scale can be seen in changes in global biomass loss. From at least as far back as 1700 until the 1960s land conversion for agriculture was the largest source of global biomass loss, but since that decade wood extraction has become larger as it grew while agricultural land expansion slowed (Hurtt *et al.*, 2006).

Over time, additional demands for the management of non-wood values such as water catchment, health, recreation, biodiversity and climate stabilisation have led to pressure for change in the forest stewardship basin. This basin becomes a place of growing tension between wood production and these new demands. In part, these tensions have been addressed by re-assigning areas of forest from wood production to other land tenures (for example, national parks). Resources are shifted into managing forests for recreation and conservation. Reflecting this, the global protected area estate grew exponentially over the twentieth century (Ervin, 2003). While these shifts are occurring in the allocation of extensive forests, there is also a shift to intensified wood production. This is reflected in the growth of the extent of plantations and trees not from forests across the globe that is described in the previous chapter (and with further empirical analysis in the next chapter).

The broad pattern of change as reflected in land use patterns raises questions of why does this occur? The following is some consideration of the economic, technological and social drivers behind these changes. Often these drivers are interconnected and influence one another. The process that is separating wood production into specialised wood cultivation segregated from forest ecosystem management is a logical outcome of optimising the competing demands on forests (Vincent and Binkley, 1993; Zhang, 2005). In addition, as noted earlier, there are economic incentives to reduce the costs of wood and wood products. Wood plantations offer a number of economic advantages in wood production and processing, such as log size consistency, greater production concentration near processing plants and much more efficient use of land.

Growing demand for non-wood values from forests act as a source of pressure in the stewardship forestry basin by also forcing innovation. Binkley (2003, p. 4) says, 'there is no limit on demand for environmental services' and, compared to wood production, 'few, if any, technical substitutes'. This mounting demand for social and ecosystem values from forests, along with other land use pressures resulting from population and economic growth, raises land costs and pushes technological and capital substitution towards intensified wood cultivation (Binkley, 2003). Further, the productivity of extensively managed forest areas is constrained by increasing environmental regulation (Binkley *et al.*, 2005, pp. 62–63) – as well as by slow natural forest regeneration rates and reductions in productivity following first harvests. Though natural forests can have a cost advantage through the lack of establishment and growing costs (Oliver and Mesznik, 2006), productivity from wood plantations typically outstrips that of natural forests by an order of magnitude (Paquette and Messier, 2009), and productivity rates in cultivated wood sources are constantly growing (Gonçalves *et al.*, 2013; O'Hehir and Nambiar, 2010).

These patterns are replicated around the world as markets, institutions and technologies are shared across jurisdictions – feedback effects across the panarchy. Together, this induced innovation, both in how wood is used to make wood products and in how it is grown and harvested, create a source of tension within wood and forest social-ecological system basins. In economics, such processes are well understood to disrupt (transform) economies out of equilibrium (resilient or stable states) (see for example, Boulding, 1981; Nelson and Winter, 1982).

However, it is also important to note that technology can play a role in system stability (resilience or resistance) as an attractor in the system, especially where complex and expensive new technologies require capital intensive investment (Hyde, 2012) or have associated resistant institutional arrangements such as distributional coalitions. The role of actors and their institutions are central to social-ecological system resilience and transformability. A key to considering these institutions is to look at the underlying rationales and social values driving them.

For forestry, the 'underlying logic' of sustainable wood production is essentially economic (Nelson, 2013). It was established to manage 'forests and other ecosystems to dampen disturbance cycles to generate a predictable and stable supply of services' (Kant *et al.*, 2013, p. 5). The stewardship forestry basin has been underpinned by the concept of sustainable yield and its inferred condition of an optimal state. However, delivering this is challenged when other levels of the panarchy, or state variables of the wood and forest social-ecological system, continually change (for example, social values or technologies). This can create a gap between the forester's models of stable equilibrium, biological systems and the directional and evolutionary nature of changes to these other dimensions of the wood and forest social-ecological system (Kant, 2000). Norgaard (2010) notes that the emphasis on stock-and-flow optimising approaches in such stable systems thinking can lead to a blindness to these other complexities.

This notion of institutional 'blindness' in forestry thinking has been described as having a religious quality (Duerr and Duerr, 1975; Nelson, 2013). Nelson (2013) refers to the underlying 'philosophy' of forestry thinking – a widespread belief in the powers of science to guide rational management of resources for the betterment of humans that emerged in the late nineteenth and early twentieth century. Nelson argues that as this belief system began to be challenged in the second half of the twentieth century, a new set of beliefs took hold within which nature was seen to have a value independent of its utility. The perspective expressed itself in the wilderness inspired conservation movement that in turn evolved into the ecology/biodiversity conservation movements. This new set of beliefs impacted on the conduct of foresters and forestry institutions (in the stewardship forestry basin), and also saw the emergence of the field of biodiversity conservation (the forest ecosystem management basin).

Overall then, two distinctive and evolving pressures are operating on the stewardship forestry basin, acting to increase the basin's precariousness and setting up the need for adaptation or transformation. First are the economic pressures pushing for ever cheaper and more efficient wood use. Second is the global shift toward increased demand for non-wood values from forests. The social drivers for this shift are complex and multifaceted, although clearly of growing international significance (Dunlap and York, 2008). The constitution of 'sustainable yield' across extensively managed areas of natural forest in the face of these changing contexts is difficult to achieve (Laband, 2013). These pressures increase the complexity of forest management in the stewardship forestry basin as they continually bring new and more demanding actors into the space (Kant *et al.*, 2013).

To some extent these tensions are reflected in shifts within the institutions of forestry. Approval of the 'doctrine of timber primacy' amongst US foresters declined over the course of the first seven decades of the twentieth century (Duerr and Duerr, 1975). This occurred with a shift of forestry focus from a pure wood supply focus to multiple-use and ecosystem service delivery (Nelson, 2013). There have been declines in enrolments in forestry courses in recent decades in many countries (Leslie *et al.*, 2006) due to perceptions that forestry is an industry in decline; negative views developed through public conflicts over forestry; and shifting of educational resources to training land managers focused on forest ecosystem management rather than wood production (Ferguson, 2012). Kennedy *et al.* (2001. p. 93) argue that in response to the transition of rural economies in industrial societies to 'urban, post-industrial global societies' there is a need for an evolution of 'foresters managing forests for the public to … natural resource professionals who manage public forest ecosystems with the people'.

These pressures on the wood and forest social-ecological system create tensions between taking approaches of adaptation or transformation. The limitations of old approaches of adaptation are illustrated in attempts to define *forests*, *forestry* and *foresters*. In an influential paper addressing the need for clearer definitions of forest types, Carle and Holmgren (2003, p. 2) state: '[t]he broad agreement from recent definitions processes is that "Forests" are tree covered areas not predominantly used for purposes other than forestry.' So, *forests* is then defined by the presence of the social institutions of *forestry*. The act of definition here is more about staking a conceptual claim for the legitimacy of *foresters* and *forestry* than it is about forests, trees or even forest ecosystems per se. This conceptual confusion can also be seen when Vanclay (2007, p. 885) says the *forester* 'manages ecosystems characterised by trees', and that what distinguishes *foresters* from *agricultural scientists* is that the latter 'tend to manage resources at paddock scale for an annual production cycle'. By this logic agricultural scientists who contend with perennial crops and cropping systems dominated by trees (such as fruit orchards, viticulture and tropical agroforestry crops such as spices, rubber and palm oil) would be *foresters*, but it is doubtful whether this is what Vanclay had in mind.

These semantic confusions are important because they highlight the breakdown in the underlying logic in the stewardship forestry basin – the wood-forest nexus. This points to a need for institutional transformation. Such shifts are not without precedent – Peluso and Vandergeest (2001, p. 769) observe that in late colonial Southeast Asia, rubber, quinine and coffee were 'removed from the foresters' jurisdiction', and subsequently 'defined as "agricultural"'. What is at stake here is fundamental. If wood production becomes essentially agricultural and forest ecosystem management becomes devoid of wood production what needs to be done with the institutions of forestry? If the driver and logic of the institution of forestry was wood from forests, do the central institutions of stewardship forestry, including the role of the professional forester continue to have relevance? The transformational nature of the change in which wood production and forest ecosystem management diverge clearly support the idea of what could be called post-forestry conditions. These post-forestry conditions are captured in the two states of wood cultivation and forest ecosystem management.

Panarchy effects

Now that the basic model has been described and a detailed look taken at some of the key change pressures within the basins, it would be useful to consider some of the panarchy effects. In particular, how different wood and forest social-ecological systems interact with one another. A significant consideration here is the timing of wood and forest social-ecological system evolution in relation to three global phenomena: the emergence of a world system and its developmental stages; the rise in the late twentieth century of an ecological consciousness and a pursuit of non-wood values in forests; and the level of technology available to a wood and forest social-ecological system at the time of other key changes in the social-ecological system. Also discussed are the role of national boundaries and the effects of globalised trade as additional panarchy influences on wood and forest social-ecological systems.

Different wood and forest social-ecological systems undergo change at different rates and times. A country's wealth and development status, the overall role of forestry in its economy, levels of urbanisation, the relation of society to its forests and mixes of indigenous/invader societies can all influence basin resilience and transformation responses. Countries and regions are different in many of these regards. However, this all occurs in the context of an increasingly interconnected world. This means that different wood and forest social-ecological systems will be subject to distinctive panarchy influence from outside. This can be thought of particularly when wood and forest social-ecological systems are considered on national (and regional) scales, and when they are considered in relation to global historical phenomenon.

Many nations have been subject to unique historical circumstances and even shared historical experiences such as patterns of industrialisation; social development can effect nations differently depending in when the changes occurred in relation to other globalised patterns of change. As an example, the forests of Europe were significantly affected by the demand for wood to build the growing cities and navies that were required during colonial expansion and industrialisation. By comparison, the extensive forests of Indonesia's outer islands remained untouched until the later decades of the twentieth century due to distance from markets and a lack of suitable technologies to access the wood, cut it and shift it to market. Technological, political, economic and social conditions, both within the system and contextually (the panarchy), were different. While Indonesia did create (teak) plantation forestry in the sixteenth century – before that of Germany – Germany had a unique historical context that allowed its forestry to spread to global dominance, even being imported into Indonesia in the nineteenth and twentieth century through processes of colonialism and post-colonial international development. Using this example further, in the early twenty-first century Indonesia began growing acacia and eucalypt plantations which are subject to the influence of a global pool of expertise that was not available in the sixteenth century when teak plantings were being developed in Java. Such variation, in the timing of panarchy influences on national social-ecological systems intersect with the distinct features and historical conditions of any other system, contributes to distinctive differences in how the model of the transition occurs.

The differences in how this model has played out in Europe, as opposed to much of the rest of the world, demands further consideration here. Many national/colonial wood and forest social-ecological systems were subject to the importation of European ideas about forest management. As noted earlier, these ideas were initially developed in parts of Europe after the exploitation of their forests was almost total, in a time when there was little or no social demand for biodiversity, conservation or ecological sustainability. Further, the 'forests' that the original German foresters managed were largely forests of their own planting (Evans, 2009). However, European silvicultural practices were developed around the growing of new forests as collections of trees designated primarily for wood production. This was a quite different logic to the subsequent application of stewardship forestry thinking to the extant natural forest ecosystems of colonised parts of the world.

Of course, this process is not just about dissemination of knowledge, institutions and technologies – it is also shaped by political power. It has already been noted how stewardship forestry has been deployed as tool of territorial control in places such as India, Indonesia,[1] the US and others. Further, in the post-colonial period the dissemination of knowledge, institutions and technologies have continued by developed countries exporting forestry expertise to developing countries through the guise of international development (Vandergeest and Peluso, 2006). And these panarchy processes of international influence are not restricted to the stewardship forestry basin. Such processes are also documented in the post-forestry basins. Environmental awareness of forest use systems has been particularly influential in developed countries, from where it extended globally. But, some authors have noted the same post-colonial tendencies in the pushing of environmental protection in relation to forests (Agarwal and Narain, 1991; Nelson, 2003; Wright, 2011), and wood cultivation (Casson, 2004). However, as well as being an extension of the developed core's processes of colonisation, such spreading of ideas also reflects an emerging global environmental awareness within which it is understood that actions in one social-ecological system have consequences for other systems. Ecological thinking's insight of systemic interconnectedness carries its own logic for action and influence beyond one's own social-ecological system. But whether the spread of such ideas is power related or emerging from new knowledge of interconnectedness, the point remains that wood and forest social-ecological systems are subject to critical panarchy influences of knowledge from outside systems.

Another significant form of panarchy influence is technology. Technological change has the effect of constantly altering system dynamics. This acts both within systems and also acts through the transfer of technology from one system to another. For the stewardship forestry basin this is important. This process is significantly at odds with stewardship forestry's strong emphasis on system equilibrium – where change itself is likely to be approached as cyclical. The preceding environmental history section highlighted how the socio-economic nature of wood production has been driven by evolving technological forces. Changes are occurring that are irreversible and directional rather than static or cyclic. Technology is important in such system change (noting that, as well as a driver for change, it can also be a factor in path dependence and lock-in).

New technologies (like social institutions as noted above) can be transferred to other places, influencing other systems in new and unplanned ways, even though those other systems might not have the conditions necessary to induce that particular technological innovation in the first place. In addition, changes resulting from these technologies are unlikely to allow systems to easily change back to a previous state. As well as an agent for change in socio-ecological systems, new technologies can involve the establishment of considerable physical, institutional and financial infrastructure to support and utilise the technologies. With these investments, significant system inertia or 'lock-in' for the new conditions comes into being. For example, a large investment in pulp production facilities can 'lock-in' a dependent wood plantation landscape in the pulp mill's hinterland.

Technology as applied to wood harvest, processing and cultivation is also influenced by technologies developed in other fields and other social-ecological systems. For example, chainsaws and motorised haulage vehicles in wood harvest were, in significant part, made possible due to technologies developed in areas unrelated to wood production or forests. However, these have acted on the wood and forest social-ecological system in profound ways. Similarly, the development of screens and electronic data storage systems in information technologies has had a significant effect on global paper demand in recent decades and, hence, wood and forest social-ecological systems, despite its development being largely driven by factors outside the wood and forest social-ecological system. An increasingly integrated global society allows for rapid cross-fertilisation of such technological ideas and impacts.

While the panarchy influences noted above are highly mobile and globally ubiquitous, it is useful to remember that nation states have distinctive political, cultural and biophysical conditions that interact with these panarchy influences. The political conditions in particular are strongly defined by nation states as core units of power in the world system. It is in this context that the logic and aspiration of national self-sufficiency in wood production emerges. Nation states have boundaries that are enforced and rigid, with real world influences on social-ecological systems on either side of the boundary. For nation states, sovereignty and security are key concerns. Not being beholden to another nation's wood supplies and having some certainty of supply of wood, a material that has often been seen as central to economic development, is important. This can push countries that might otherwise have been able to more efficiently import wood by playing to their comparative advantage, to instead continue to invest in existing wood and forest social-ecological systems within their boundaries to achieve greater wood supply self-sufficiency. This can act as a counter balance to the panarchy influences of external values, knowledge, institutions and technologies noted above.

This counterbalance to panarchy pressures on basins can be reinforced within national (and regional) systems as discourse coalitions and distributional coalitions within wood forest social-ecological systems act to maintain a status quo (for example, timber industry resistance to increased forest conservation or demands for subsidies from governments to counter global market and technological pressures). This can act to lock in a social-ecological system in certain jurisdictions, giving greater resilience to the system in the face of these other pressures. However, the logic of national independence (and related political needs to perpetuate local wood dependent economies) can drive countries to subsidise wood harvest from natural forests to unsustainable levels, pushing stewardship forestry systems to crises. This unsustainable use of natural forests leads governments to wood cultivation. And, as has already been noted, the drive to self-sufficiency has also acted as a pressure within the system to shift industry into the wood cultivation basin, especially through increased plantation establishment. A notable contrary example to these processes is Japan, a country that through the second half of the twentieth century largely imported wood from other countries while growing a large national forest estate that was protected from wood production for non-wood values (Dauvergne, 1997). Thus the country made decisions to outsource wood production even when its own forests had the potential to supply more wood – in distinctive contrast to the examples of nations pursuing self-sufficiency in wood production even when it is not efficient to do so.

Large countries such as Australia, the United States, Brazil and Indonesia, with strong sub-regional polities, are also likely to encompass a range of the basins described here as occurring in different sub-regions. For example, Brazil simultaneously has a major wood cultivation social-ecological system operating in the Atlantic south east that is based on a history of extensive forest exploitation in an earlier time period, while still having frontier wood exploitation occurring in the Amazonian north. The United States has seen significant shifts in wood production from the natural forests of the Pacific Northwest to plantations in the south east of the country in the last few decades. This nested system quality of large nation states is an important caveat to bear in mind when considering the application of the model at the level of the nation state.

Finally, there are of course distinctive biophysical factors that affect these processes of change. For wood and forest social-ecological systems there is considerable variety in how forests respond to wood extraction, based on the physiology of the trees, climate, soil and evolutionary history (including exposure to humans). For example, the forests of New Zealand, while producing some highly valued wood in the early stages of forest exploitation, were dominated by slow growing species. These never generated growth rates capable of producing large quantities of wood on an ongoing basis. This in turn contributed to New Zealand rapidly shifting to wood cultivation. Likewise, the tropical rainforests of the world have largely proved to be slow growing and low yielding. This is in contrast to the pine species of the temperate and boreal forests of the northern hemisphere and the eucalypt species of Oceania. Temperate forests are dominated by fewer and similar species where the logging tends to a clear-cut (removing most of the trees in a single encroachment). Tropical forests on the other hand receive much more selective logging, often in the form of high grading of sequentially lower value species (Hyde, 2012). While these biophysical qualities indicate distinctiveness of social-ecological systems, the processes of globalisation are also interacting with these to act as powerful panarchy pressures again. For example, the tropics and subtropics are suitable for fast growing tree crops such as plantations. This has led to the shift of global wood production towards the equator. This panarchy influence can also be seen in the global spread of two specific groups of tree for wood cultivation – pines and eucalypts. Wood cultivation has been able to boost wood productivity through these species; first, by utilising the best available species for particular conditions from anywhere in the world and, second, by also bringing species into new environments where they have not occurred before and thus do not have indigenous diseases and pests to contend with, further enhancing productivity.

The discussion above outlines a number of pervasive global panarchy influences (political, social and technological) on the world's wood and forest social-ecological systems. It focused particularly on the basin of stewardship forestry and the pressures acting on it and potential adaptive and transformational responses. It also notes a number of countervailing processes, including organised social resistance and distinctive developmental, historical and biophysical conditions that influence the basins, and change between basins within any one social-ecological system. These are important considerations to the model. They provide caution for the models' application, but also build an understanding of the process that the book will come back to in the case studies of Part II, and the concluding discussion and analysis in Part III.

Note

1 For a detailed analysis on Indonesia, see the later chapter *A social turn: decentralisation in Indonesia – a forestry history*.

References

Agarwal, A. and Narain, S. (1991). Global warming in an unequal world: a case of environmental colonialism. *New Delhi, Centre for Science and Environment.*

Binkley, C.S. (2003). Forestry in the long sweep of history. In: Teeter, L.D., Cashore, B.W. and Zhang, D. (eds.) *Forest policy for private forestry: global and regional challenges*. Oxon: CABI.

Binkley, C.S., Aronow, M.E., Washburn, C.L. and New, D. (2005). Global perspectives on intensively managed plantations: implications for the Pacific Northwest. *Journal of Forestry*, 103, 61–64.

Boulding, K.E. (1981). *Evolutionary economics*. Beverly Hills, CA: Sage Publications.

Carle, J. and Holmgren, P. (2003). Definitions related to planted forests. *UNFF intersessional expert meeting international steering group on "the role of planted forests in sustainable forest management. Maximising planted forests' contribution to SFM"*. Wellington.

Carpenter, S., Walker, B., Anderies, J.M. and Abel, N. (2001). From metaphor to measurement: resilience of what to what? *Ecosystems*, 4, 765–781.

Casson, A. (2004). The controversy surrounding eucalypts in social forestry programs of Asia. *Canberra: Economics Division, Research School of Pacific and Asian Studies, Australian National University*.

Dauvergne, P. (1997). *Shadows in the forest: Japan and the politics of timber in Southeast Asia*. Cambridge, MA: MIT Press.

David, P.A. (1985). Clio and the economics of QWERTY. *The American Economic Review*, 75, 332–337.

Diamond, J.M. (2005). *Collapse: how societies choose to fail or succeed: revised edition*. London: Penguin UK.

Duerr, W.A. and Duerr, J.B. (1975). Role of faith in forest resource management. In: Rumsey, F. and Duerr, W.A. (eds.) *Social sciences in forestry: a book of readings*. Philadelphia, PA: Saunders.

Dunlap, R.E. and York, R. (2008). The globalization of environmental concern and the limits of the post-materialist values explanation: evidence from four multinational surveys. *The Sociological Quarterly*, 49, 529–563.

Ervin, J. (2003). Protected area assessments in perspective. *BioScience*, 53, 819–822.

Evans, J. (2009). *Planted forests: uses, impacts and sustainability*. Rome: CABI International and FAO.

Ferguson, I. (2012). Future forestry employment and education. *Australian Forestry*, 75, 192–199.

Folke, C., Carpenter, S., Elmqvist, T., Gunderson, L., Holling, C.S. and Walker, B. (2002). Resilience and sustainable development: building adaptive capacity in a world of transformations. *AMBIO: A Journal of the Human Environment*, 31, 437–440.

Gonçalves, J.L.D.M., Alvares, C.A., Higa, A.R., Silva, L.D., Alfenas, A.C., Stahl, J., Ferraz, S.F.D.B., Lima, W.D.P., Brancalion, P.H.S. and Hubner, A. (2013). Integrating genetic and silvicultural strategies to minimize abiotic and biotic constraints in Brazilian eucalypt plantations. *Forest Ecology and Management*, 301, 6–27.

Hajer, M. (1993). Discourse coalitions and the institutionalization of practice: the case of acid rain in Britain. In: Fischer, F. and Forester, J. (eds.) *The argumentative turn in policy analysis and planning*. Durham and London: Duke University Press.

Hardin, G. (1968). The tragedy of the commons. *Science*, 162, 1243–1248.

Herrfahrdt-Pähle, E. and Pahl-Wostl, C. (2012). Continuity and change in social-ecological systems: the role of institutional resilience. *Ecology and Society*, 17, 8.

Hodgson, G.M. (1993). *Economics and evolution: bringing life back into economics*. Cambridge: Polity Press.

Holling, C.S. (1973). Resilience and stability of ecological systems. *Annual Review of Ecology and Systematics*, 4, 1–23.

Hurtt, G., Frolking, S., Fearon, M., Moore, B., Shevliakova, E., Malyshev, S., Pacala, S. and Houghton, R. (2006). The underpinnings of land-use history: three centuries of global gridded land-use transitions, wood-harvest activity, and resulting secondary lands. *Global Change Biology*, 12, 1208–1229.

Hyde, W.F. (2012). The global economics of forestry. *New York, Resources for the Future*.

Kant, S. (2000). A dynamic approach to forest regimes in developing economies. *Ecological Economics*, 32, 287–300.

Kant, S., Wang, S., Deegen, P., Hostettler, M., Von Detten, R., Howard, T., Laband, D., Montgomery, C., Robert, N. and Sekot, W. (2013). New frontiers of forest economics. *Forest Policy and Economics*, 35, 1–8.

Kennedy, J.J., Thomas, J.W. and Glueck, P. (2001). Evolving forestry and rural development beliefs at midpoint and close of the 20th century. *Forest Policy and Economics*, 3, 81–95.

Laband, D.N. (2013). Public choice, rent-seeking and the forest economics-policy nexus. In: Kant, S. (ed.) *Post-Faustmann forest resource economics*. Dordrecht, The Netherlands: Springer.

Leslie, A., Wilson, E. and Starr, C. (2006). The current state of professional forestry education in the United Kingdom. *International Forestry Review*, 8, 339–349.

MacIlwain, C. (2014). Emerging powers need a more-inclusive science. *Nature*, 505, 7.

Matthews, E. (1983). Global vegetation and land use: new high-resolution data bases for climate studies. *Journal of Climate and Applied Meteorology*, 22, 474–487.

Nelson, R.H. (2003). Environmental colonialism: "saving" Africa from Africans. *The Independent Review*, 8, 65–86.

Nelson, R.H. (2013). Multiple-use forest management versus ecosystem forest management: a religious question? *Forest Policy and Economics*, 35, 9–20.

Nelson, R.R. and Winter, S.G. (1982). *An evolutionary theory of economic change*. Cambridge, MA: Belknap.

Norgaard, R.B. (2010). Ecosystem services: from eye-opening metaphor to complexity blinder. *Ecological Economics*, 69, 1219–1227.

O'Hehir, J. and Nambiar, E. (2010). Productivity of three successive rotations of P. radiata plantations in South Australia over a century. *Forest Ecology and Management*, 259, 1857–1869.

Oliver, C.D. and Mesznik, R. (2006). Investing in forestry: opportunities and pitfalls of intensive plantations and other alternatives. *Journal of Sustainable Forestry*, 21, 97–111.

Olson, M. (1982). *The rise and decline of nations: economic growth, stagflation and social rigidities*. New Haven, CT: Yale University Press.

Paquette, A. and Messier, C. (2009). The role of plantations in managing the world's forests in the Anthropocene. *Frontiers in Ecology and the Environment*, 8, 27–34.

Peluso, N.L. and Vandergeest, P. (2001). Genealogies of the political forest and customary rights in Indonesia, Malaysia, and Thailand. *Journal of Asian Studies*, 60, 761–812.

Pierson, P. (2000a). Increasing returns, path dependence, and the study of politics. *American Political Science Review*, 95 (2) 251–267.

Pierson, P. (2000b). The limits of design: explaining institutional origins and change. *Governance*, 13, 475–499.

Pongratz, J., Reick, C., Raddatz, T. and Claussen, M. (2008). A reconstruction of global agricultural areas and land cover for the last millennium. *Global Biogeochemical Cycles*, 22.

Rist, L. and Moen, J. (2013). Sustainability in forest management and a new role for resilience thinking. *Forest Ecology and Management*, 310, 416–427.

Vanclay, J.K. (2007). Educating Australian foresters for the 21st century. *International Forestry Review*, 9, 884–891.

Vandergeest, P. and Peluso, N.L. (2006). Empires of forestry: Professional forestry and state power in Southeast Asia, Part 2. *Environment and History*, 12, 359–393.

Vincent, J.R. and Binkley, C.S. (1993). Efficient multiple-use forestry may require land-use specialization. *Land Economics*, 69, 370–376.

Walker, B., Holling, C.S., Carpenter, S.R. and Kinzig, A. (2004). Resilience, adaptability and transformability in social–ecological systems. *Ecology and society*, 9, 5.

Walker, J. and Cooper, M. (2011). Genealogies of resilience. From systems ecology to the political economy of crisis adaptation. *Security Dialogue*, 42, 143–160.

Williams, M. (2002). *Deforesting the earth: From prehistory to global crisis*. Chicago: University of Chicago Press.

Wright, G. (2011). Indigenous people and customary land ownership under domestic REDD+ frameworks: a case study of Indonesia. *Law Environment and Development Journal*, 7, 119–131.

Young, O.R. (2010). Institutional dynamics: resilience, vulnerability and adaptation in environmental and resource regimes. *Global Environmental Change*, 20, 378–385.

Zhang, Y. (2005). Multiple-use forestry vs. forestland-use specialization revisited. *Forest Policy and Economics*, 7, 143–156.

Part II

Case studies

Countries, regions and themes

4 Global analysis of trends in wood sourcing

This chapter seeks to define the extent to which global wood sourcing transition is occurring. An effective understanding of what is taking place needs a global perspective. There are two main reasons for this. First, as already noted, nation states are at different stages of economic development and have widely varying conditions. At the level of individual states (nations and states within nations) there is a wide range of conditions of forests, their extent and their use for wood extraction. Choosing to look at any one state or a handful of states is likely to provide a limited picture. Second, the social and economic systems that are operating in each nation interdepend on those occurring within others. Indeed, many aspects of the economy of wood and forest use are international, not to mention such interconnected biophysical systems as water, biodiversity and carbon that operate at this scale. For this reason a global analysis is effective in bounding a system in a way that is more difficult at smaller scales.

A key challenge in past attempts to quantify the international picture of plantation forest trends has been the inconsistency of data. In a review by ABARE (1999, p. 80) the authors noted that, 'the information available on plantations reflects more on what is *not* known about plantations.' This has been possibly further confused due to the definitional issues noted by Carle and Holmgren (2003). The research presented in this chapter tackles this problem and seeks to clarify the magnitude of the pattern of change in wood sourcing from natural forests to cultivated wood.

While a large portion of the world's wood is produced and consumed within national boundaries, international trade has a significant impact on price, production and policy within these countries. Further, global power is influenced by numerous other factors, such as international treaties, the activities of non-government organisations, international third party certification and globally organised capital – particularly in the form of transnational corporations. These transnational and supranational influences are vital to understanding global processes, but also critical to understanding the peculiarities of national or local conditions. The reach of global processes is never far away and, thus, is a critical consideration for all with an interest in wood production and forest management. It is within this global perspective that the transition of wood production is analysed here.

Introduction

Wood has been a vital natural resource for humans. It has been the primary source of fuel for most of human existence and central to building, tool making, information sharing (in the form of paper) and shipping. Mostly, wood has been obtained from naturally regenerated forests, much as food was primarily collected from natural sources before the advent of agriculture. However, in response to local or national wood shortages there has been a move to intentional wood cultivation, essentially as crops.

The decline in natural forest wood extraction in many nations is consistent with the 'Hubbert Curve', a rapid increase in production to a peak followed by a sharp initial decline, followed by a slow tapering off (Hubbert, 1993). Shearman *et al.* (2012) demonstrate that 'peak wood' has been passed in the Solomon Islands, the Philippines, Thailand, Malaysia, Laos and Indonesia. Countries supplying a large portion of China's wood in 2006 were harvesting at unsustainable rates, with only years to decades left to economically access mature natural forest: in the Russian far east it was 20 years; in Indonesia, ten years; Papua New Guinea, 13–16 years; Myanmar, 10–15 years; Cambodia, 4–9 years (White *et al.*, 2006). Roundwood production in the second rotation in tropical forests was 54 per cent lower than the first cutting (Putz *et al.*, 2012). Canada's historical industrial roundwood production, largely sourced from natural forests (Kanowski and Murray, 2008), peaked in 2004 at 205 million m^3 and in 2016 was 157 million m^3 (FAOSTAT, 2018). Australia's natural forest wood production has declined 56 per cent, from 9.6 million m^3 in 1996–97 to just under 4.3 million m^3 in 2016–2017.

This chapter will assess the global trend in wood extraction from natural forests to understand how this will affect the future roles of wood production, biodiversity conservation and other ecosystem services in natural forests. While these roles are strongly influenced by regional and national factors, it is also valuable to understand how any specific local pattern of forest use is influenced by global trends, as well as identifying the likely future trajectories of local and national patterns of forest use that are derived from global patterns. This will be of particular value where wood production competes with other forest use values, and policy makers and managers have needed to 'balance' these competing demands. Finally, global trends also provide data against which the effectiveness of global forest policy and governance initiatives and structures can be measured.

Methods

In order to assess the global trend in roundwood[1] production from natural forests, three key steps were required. First, an estimation of total roundwood production was developed from historical data, along with three outlooks: a high growth projection; a projection based on best-fit trends of the historical pattern; and an intermediate pattern. Second, cultivated roundwood production estimates for the same period were developed, creating high and low bound scenarios for the two different cultivated wood groups described in pages 71–74. The third step was to subtract the cultivated roundwood estimate from total roundwood to generate natural forest roundwood production totals. This was done for each combination of total roundwood output and cultivated roundwood output variants to produce four different natural forest roundwood datasets for the historical period and, reflecting the higher levels of uncertainty, 12 for the forecast period.

The rationale for this method stems from the lack of data on wood extraction[2] from natural forest sources. The Food and Agriculture Organisation of the United Nations (FAO) global datasets on roundwood production include all sources of roundwood: natural forests, plantations, planted forests and trees not from forests. There have also been some analyses of global totals and trends in roundwood from cultivated wood sources (ABARE, 1999; Brown, 2000; Carle and Holmgren, 2008; Smeets and Faaij, 2007).[3] Between them – the FAO total roundwood data and cultivated wood analyses – they offer the opportunity to subtract one from the other to derive totals and trends for natural forest roundwood extraction.

For this analysis, the historical period back to 1945 was assessed, based on the availability of suitable data. The FAO was established in 1945 and global data prior to this period is scant. Projections to 2030 were used as this is a date that occurs frequently in published roundwood production projections.

A number of definitions are important to this analysis: roundwood types and the various roundwood sources, including natural forests, planted forests and plantations. Roundwood is an aggregate of two main subgroups: fuelwood (for energy production, industrial or domestic) and industrial roundwood (all other roundwood types such as pulplogs, sawlogs and veneer logs) (FAO, 2010a). These two roundwood types are a key categorisation employed in the FAOs statistics and other sources used.

Definitions of forest and non-forest types draw on work developed by Carle and Holmgren (2003), as discussed in the introduction, as part of attempts to understand global trends in cultivated wood development. The definitions developed (see Figure 4.1) have been widely adopted and used, most notably by the FAO. They are used in this analysis to define *natural forest* as those forest types at the natural and semi-natural end of the spectrum, distinct from roundwood sources on the other side of the spectrum, described as *cultivated wood*. The analysis will use two different sets of cultivated wood, reflecting the data available: one including planted forests and one with only plantations and trees outside forests.

The following two sub-sections detail the methods and sensitivity analysis applied to the development of the global total roundwood production figures and the global total cultivated roundwood production variants.

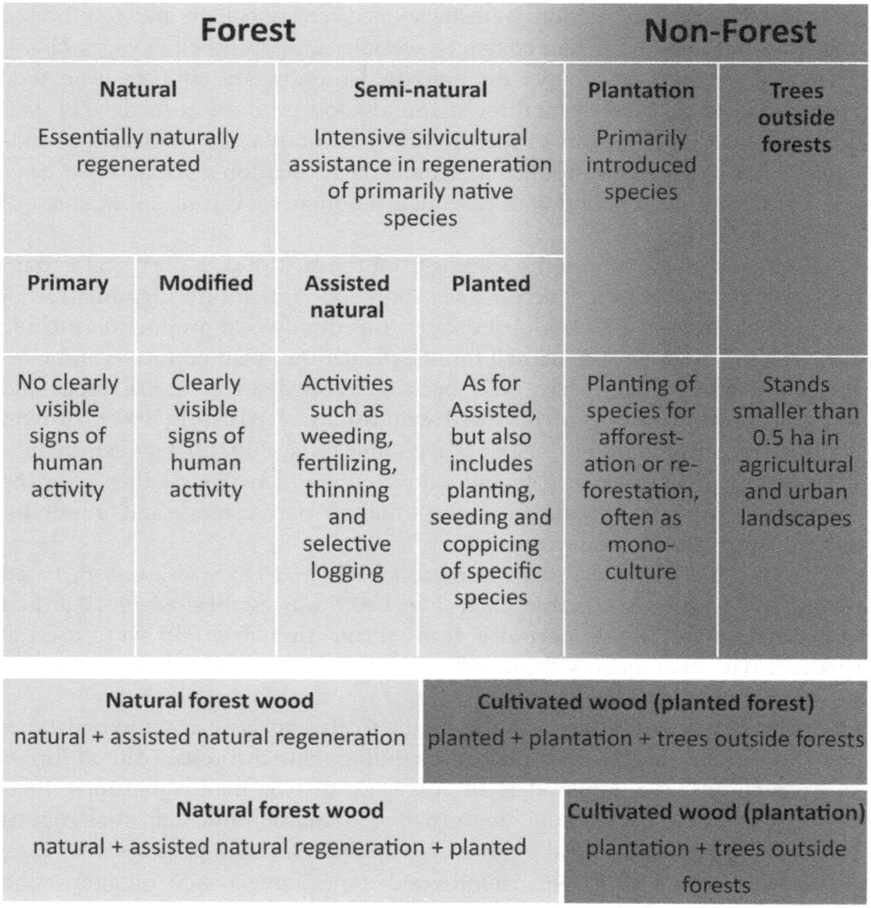

Figure 4.1 Continuum of wood source characteristics and the two corresponding sets of cultivated wood used here.

Source: modified from Carle and Holmgren (2008).

Total global roundwood production: data sources and sensitivity analysis

A historical global roundwood dataset for 1945–2016 was developed using data from the FAO. Three future roundwood production outlooks were created: a high, middle and low level of projected future roundwood production (see Table 4.1).

All roundwood data (fuel and industrial) for the period 1961 to 2016 were taken from (FAOSTAT, 2018). Suitability of the FAOSTAT data for use in

Table 4.1 Summary of total roundwood production data methods and materials

Total roundwood consumption	*Industrial roundwood*	*Fuelwood*
Historical record FAO	(FAOSTAT, 2018) for 1961–2016, FAO Yearbook of Forest Products Statistics 1962 and 1958 (FAO, 1958; FAO, 1962) for 1948–1960, and linear extrapolation for 1945–1947 from the 1948–1960 results	(FAOSTAT, 2018) for period 1961–2016 and extrapolation based on 1961–2016 trend in per capita fuelwood consumption (population data – FAOSTAT, 2018) for 1945–1960
Scenario 1 Historical trend	Extrapolation of best fit trend of 1945–2016 industrial roundwood for 2017–2030	Extrapolation based on 1961–2016 trend in per capita fuelwood consumption (population data – FAOSTAT, 2018) for 2017–2030
Scenario 2 Intermediate	Midway point for each year between Scenario 1 and Scenario 3	
Scenario 3 High Growth	Based on FAO (2009) projections for period (1.8% annual growth 2005–2020 and 1.3% subsequent), applied from 2017 onwards	1.3% annual growth from 2017 onwards based on Buongiorno et al. (2012) assuming high future demand for biofuel

this analysis was checked with the FAO Forestry Department. This data was generated from annual country surveys, 'unofficial' sources such as trade journals, industry association reports, and statistical yearbooks and FAO estimates for the remainder (including ongoing adjustments when problems such as unreported production are found). The FAO estimates for industrial roundwood are based on previous years' figures, and for fuelwood by use of methods refined following a review of fuelwood consumption in response to concerns[4] that prior estimates were inaccurate (Whiteman *et al.*, 2002). A recent internal FAO assessment found that approximately 20 per cent of the global quantity of industrial roundwood and 85 per cent of fuelwood was estimated by these methods. There have been no definition changes that would limit the comparability of the FAOSTAT series data back to 1961. Overall, the use of the data in this analysis was considered sound (Arvydas Lebedys [FAO, Forestry Department] pers. comm., 22 and 28 September 2013).

Industrial roundwood figures for the period 1948–1960 were taken from the 1958 and 1962 FAO *Yearbook of Forest Products Statistics* (FAO, 1958, 1962). Where these sources reported different figures in overlapping years, the data from the latter report were used (as advised by the FAO). The period from 1945–1947 was derived using a linear extrapolation of the period covered by the published yearbooks (1948 to 1960). The resulting figure of

605 million m^3 of industrial roundwood for 1945 is consistent with the figure of 600 million m^3 reported by Solberg *et al.* (1996) from the FAO.

Fuelwood consumption reported in annual reports (FAO, 1958, 1962) for the period prior to FAOSTAT data (that is, before 1961) is considerably lower than the FAOSTAT data would suggest. For example, the FAO *Yearbook of Forest Products Statistics 1962* (FAO, 1962) fixed global fuelwood for 1961 at 755 million m^3, with figures ranging from 792 to 696 million m^3 for the period 1951 to 1961. This is out of step with the FAOSTAT figures of 1,498 million m^3 for 1961 and only a small trend upward in subsequent years – just climbing to 1,550 million m^3 in 1971. The 1962 *Yearbook* notes that 'figures for fuelwood removals are far more incomplete than those for industrial roundwood' and that 'statistics for fuelwood removals are still considered to understate actual removals' (FAO, 1962, p. VII). Datasets for fuelwood underwent major revision in 2002 (Whiteman *et al.*, 2002). Given the FAO's Yearbook's unreliably low fuelwood estimates before 1961, a backwards extrapolation of FAOSTAT data using the exponential trend in annual global per capita fuelwood consumption for the period 1961 to 2016 ($y = 0.4781e^{-0.012x}$, $R^2 = 0.9918$) was used to provide fuelwood figures for 1945 to 1960.

Totals for global industrial roundwood and fuelwood, developed by the above methods, were then combined to derive global total roundwood production for the period 1945 to 2016.

In order to develop a likely range of future scenarios, a review of past outlooks was undertaken. In the late 1990s considerable effort went into developing global industrial roundwood forecasts. Many of these studies were subsequently found to have overestimated projected production. A failure to sufficiently account for growth in resource productivity has been suggested as a key reason for this (Ajani, 2011; Smeets and Faaij, 2007). Two reviews of published outlooks, Smeets and Faaij (2007. p. 13), supplemented and modified from Weiner and Victor (2000) and ABARE (1999), list between them a total of 37 scenarios with an outlook for 2010. These were produced by 15 authors between 1994 and 2001. The outlooks for 2010 had an annual mean of 1,961 million m^3, and ranged from 1,510 million m^3 to 2,700 million m^3. The recorded production in 2010 was 1,704 million m^3 (FAOSTAT, 2018) – only three of the 37 forecasts were lower than this. Similar patterns of overshoot in industrial roundwood projections are noted in fuelwood projections from the same sources.

Given this tendency for past published outlooks to exceed actual production, projections using best-fit trends of historical data were used to give a feasible low end scenario. In the case of industrial roundwood this was based on historical volumes, and in the case of fuelwood it was based on historical per capita consumption trends multiplied by population projections. The validity of the low end fuelwood projections used here is backed by more recent analysis suggesting that stagnant fuelwood use is likely to continue into the future (Arnold *et al.*, 2003; Mead, 2005). The high growth variation for industrial roundwood outlook here uses projected growth rates from the FAO *State of the World's Forests 2009* (FAO, 2009), applied from 2017. These rates were 1.8 per cent per annum for 2005 to 2020 and 1.3 per cent per annum for 2021 to 2030. As with the earlier projections noted above, this set of predictions (published in 2009) has so far been much higher than actual growth – industrial roundwood grew from 1,787 million m^3 in 2005 to 1,873 million m^3 in 2016 (FAOSTAT, 2018); an annual growth of 0.4 per cent – 1.8 per cent per annum is, therefore, considered a suitable, if unlikely, upper bound growth scenario. The upper bound variation for fuelwood was based on a 2030 fuelwood projection developed using the Global Forest Products Model with Intergovernmental Panel on Climate Change assumptions of 'varying but large future biofuel production' (Buongiorno *et al.*, 2012, p. 1). This is consistent with other literature in speculating that increased future energy demand or climate initiatives will raise fuelwood consumption (FAO, 2010b; Heinimö and Junginger, 2009). The highest of the three scenarios involved a fivefold increase in the growth rate observed between 1992 and 2006 and was considered unrealistic. The second highest growth scenario was chosen. This involved annual growth of 1.3 per cent. It is considered a reasonable upper bound growth scenario. While the combined industrial roundwood and fuelwood upper bound growth scenario used here is considered unlikely, it gives a conservative test on the impact of future growth on the trends in natural forest wood extraction.

Total global roundwood from cultivated wood variants: data sources and sensitivity analysis

Data for cultivated roundwood production was generated from published outlook studies and estimates of trends in roundwood production from these sources. As noted earlier, the distinction between natural forest and other sources of wood is not a neat duality but, rather, two ends of a spectrum with a number of possible points to delineate the two. In recognition of this, four variants of cultivated roundwood were developed: a high and low bound estimate based on productive plantations, as used in Brown (2000), and trees outside forests; and a high and low bound estimate based on the broader category of planted forests, as used in Carle and Holmgren (2008), and trees outside forests (Table 4.2). Both Brown (2000) and Carle and Holmgren (2008) utilised data collected through international surveys undertaken in conjunction with the FAO. Both sets of work necessarily include a range of assumptions (such as stand age of future harvests, current and future silvicultural methods, and rates of production efficiency gains) that would affect predicted outcomes if they were to change. In all variants, roundwood from trees outside forests was estimated drawing on the work of Smeets and Faaij (2007).

Brown (2000) included three scenarios. Scenario 1 was based on no new plantation establishment from 1995. This has clearly not been the case and it is not used here. Scenario 2 was chosen as a low end case. It was based on a

Table 4.2 Summary of total cultivated roundwood production data methods and materials

Variants	*Industrial roundwood and fuelwood past estimates*	*Industrial roundwood and fuelwood forward estimates*	*Fuelwood from trees outside forests past and forward estimates*
Plantation – low estimate	Brown (2000) Scenario 2 1995 estimate than roughly exponential decline in steps 1990–1995; 1970–1989; 1960–1969; 1945–1959	Brown (2000) Scenario 2, 1995–2030 projections	Assume grew from a portion of 1/10 of total fuelwood in 1945 to 1/3 in 2007 and the portion continued to grow at same rate afterwards
Plantation – high estimate	Brown (2000) Scenario 3 1995 estimate than roughly exponential decline in steps 1990–1995; 1970–1989; 1960–1969; 1945–1959	Brown (2000) Scenario 3, 1995–2030 projections	Assume grew from a portion of 1/3 of total fuelwood in 1945 to 2/3 in 2007 and the portion continued to grow at same rate afterwards
Planted forest – low estimate	Carle and Holmgren (2008) projected back from 4/5 of 2005 estimate in same linear trend as projected forward from 4/5 of 2005 estimate to Scenario 2 estimate for 2030 until 1990 and then at half that rate from 1990 backwards	Carle and Holmgren (2008) projected up from 4/5 of 2005 level in linear fashion to Scenario 2, 2030 level	Same as Plantation – low
Planted forest – high estimate	Carle and Holmgren (2008) Scenario 3 projected back from 2005 estimate in same linear trend as projected forward from 2005 until 1990 and then at half that rate from 1990 backwards	Carle and Holmgren (2008) Scenario 3, projection from 2005 to 2030	Same as Plantation – high

projected growth of plantation area of 1 per cent per annum. Del Lungo *et al.* (2006) published productive plantation figures (79 million hectares in 1990, 99 million hectares in 2000 and 111 million hectares in 2005) where annual growth rates were 2.55 per cent between 1990 and 2000 and 2.25 per cent between 2000 and 2005. This is much higher than Brown's 1 per cent for Scenario 2 consequently this scenario is used in this analysis as the low bound variant. Scenario 3 used the current planting rates of 1995 and extended them until 2004, subsequently reducing them progressively to 20 per cent of the 1995 rate of growth in 2035. This is used as a conservative high bound plantation variant.

The two planted forest variants were adapted from Carle and Holmgren (2008). Of these, the authors considered Scenario 3 to be the 'most probable scenario until 2030' (Carle and Holmgren, 2008, p. 15). It is used here as a conservative high bound variant. Scenario 2 is based on the planted forest estate growing in line with past trends, but does not factor in the productivity increases of that estate used in Scenario 3. It is used here as a low bound variant. It was modified by reducing the level of the 2005 estimate of planted forest roundwood production by four-fifths to give a lower overall planted forest roundwood output than the high bound variant for both historical and outlook periods.

The use of roughly exponential increases in roundwood outputs in the four variants is designed to replicate the historical patterns of plantation and planted forest establishment. Published accounts suggest an exponential pattern of plantation and planted forest establishment, with uptake at a relatively low level in European countries through the later centuries of the second millennia. In the first half of the twentieth century this spread to countries such as Australia, New Zealand, South Africa and the US. In the last 50 years there has been a dramatic global expansion of plantations, especially in developing countries (Evans, 2009, pp. 5–22; Varmola *et al.*, 2005). The FAO notes that 'most of the long term growth in wood supply is occurring in countries that have established planted forests during the last few decades' (FAO, 2010a, pp. 86–87). In 1995 the age class of the world's plantations was heavily skewed to those less than 15 years old, indicating the fast rate of recent expansion of global plantations (Brown, 2000). This is consistent with the findings of Del Lungo *et al.* (2006).

The FAO estimates of total roundwood (industrial and fuel) include roundwood from forests (natural and planted) and non-forest sources such as woodlands and trees not from forests. It is estimated that globally there are over one billion hectares of agricultural land with greater than 10 per cent tree cover that is not formally considered forest (Zomer *et al.*, 2009). This reserve of trees is likely to supply significant portions of the world's fuelwood supply, especially in developing countries where much of the world's fuelwood is consumed. For example, Africa's significant fuelwood consumption may come largely from trees outside forests (FAO, 2010a, p. 172). It has been estimated that two-thirds of global fuelwood consumption is from trees outside forests (Smeets and Faaij, 2007), with 'planted trees on farmland, in villages and homesteads and along roads and waterways contributing significantly to fuelwood supplies, enabling the demand to be met in most instances' (FAO, 2001, p. 23). It is likely that the amount of fuelwood coming from cultivated trees is increasing, with plantation and agroforestry planting for this purpose beginning in the 1970s on a large scale in Africa and Asia (Hyde *et al.*, 1996). Based on this, two scenarios for the contribution of trees not from forests to fuelwood production were used here. The low bound variants assume a one-third contribution to fuelwood at 2007 – half of Smeets and Faaij's (2007) assessment of two-thirds – climbing from one-tenth in 1945 (a negligible base, but assuming some limited contribution from trees outside forests). The high bound variants use a linear trended portion of the annual total, starting at one-third in 1945, increasing to Smeets and Faaij's (2007) two-thirds in 2007, and growing at the same rate beyond.

Results

The results presented start with findings for total global roundwood production, followed by findings for total global cultivated roundwood production and, finally, the results for total global natural forest roundwood production, derived by subtracting cultivated roundwood from total roundwood.

The first component of the historical roundwood production analysis was the period prior to that recorded in the FAOSTAT data (before 1961). Industrial roundwood production figures for the period 1945 to 1960 are shown in Figure 4.2.

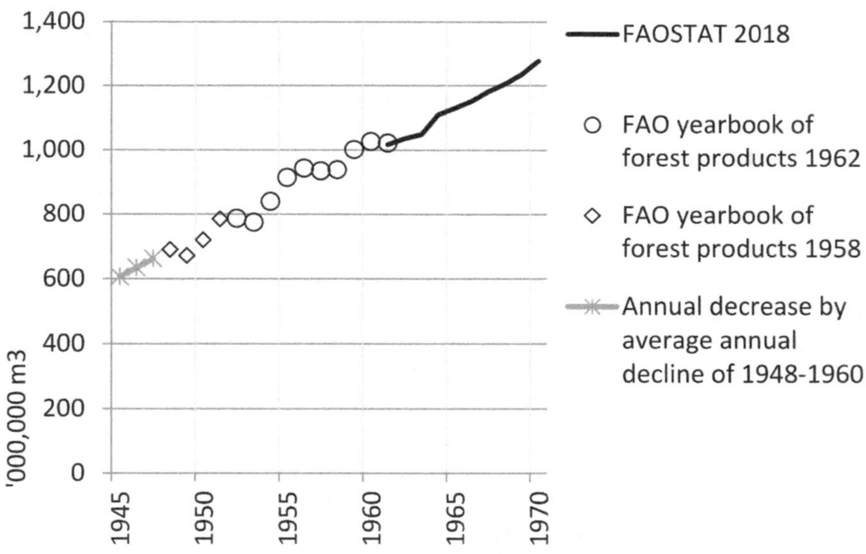

Figure 4.2 Results for total historical industrial roundwood production 1945–1970.

The total industrial roundwood data trend determined for 1945 to 1960, shown in Figure 4.2, combined with the FAOSTAT data for 1961 to 2016 generate a historical record for the period 1945 to 2016, which is shown as 'Historical record FAO' in Figure 4.3. Three scenarios of total industrial roundwood outlook for the period 2017–2030 were generated and are also shown in Figure 4.3. Scenario 1 shows a projection consistent with a best-fit

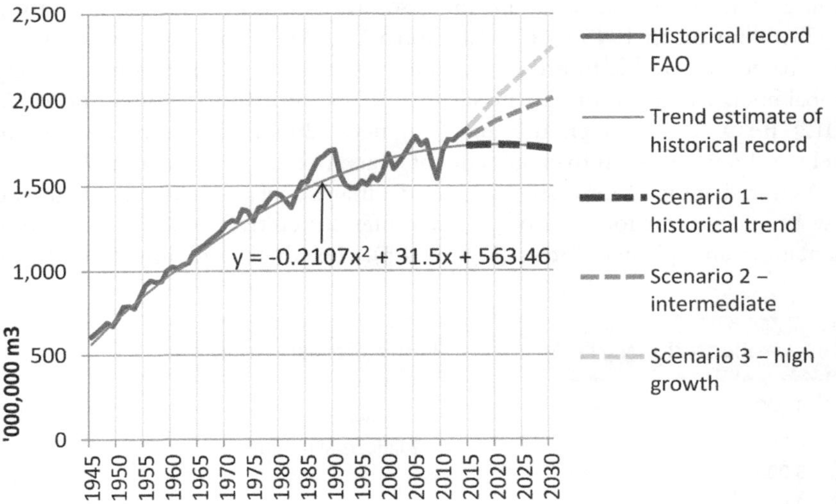

Figure 4.3 Total industrial roundwood production for historical period 1945–2016 and three forecast scenarios for period 2017–2030.

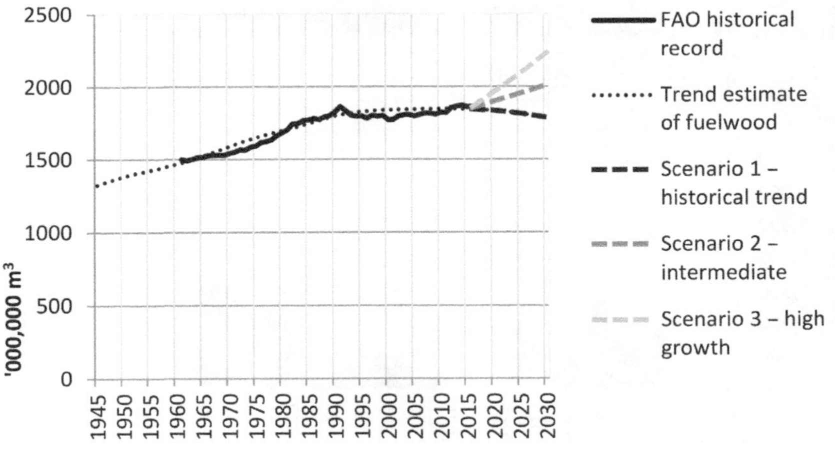

Figure 4.4 Total fuelwood production for historical period 1945–2016 and three forecast scenarios for the period 2017–2030.

trend (second degree polynomial) for the period 1945 to 2016, leading to a slow decline in total industrial roundwood production over the coming decades, reaching 1,749 million m^3 in 2030. Scenario 3, as the upper bound, gives a rate of sustained growth leading to 2,290 million m^3 total industrial roundwood in 2030.

Global fuelwood consumption per capita declined from 0.485 m^3 per annum in 1961 to 0.25 m^3 in 2016. Results of extrapolations are given in Figure 4.4. These showed an increase in fuelwood consumption from 1,325 million m^3 in 1945 to 1,468 million m^3 in 1960. The high growth scenario, based on possible future increases in biofuel demand in response to rising global energy demand or climate change policy initiatives (Buongiorno *et al.*, 2012; FAO, 2010b; Heinimö and Junginger, 2009), shows an increase in fuelwood consumption over subsequent years to 2,231 million m^3 in 2030.

As noted above, four cultivated wood variants were developed, a high and low bound variant for each of the two different cultivated wood categories: plantation and planted forest (Figure 4.5). The high and low bounds are

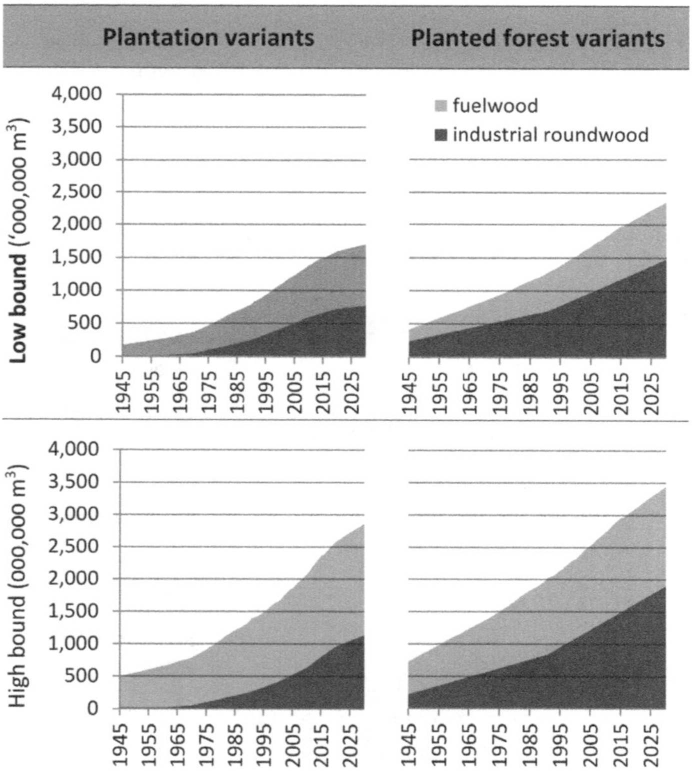

Figure 4.5 Estimated fuelwood and industrial roundwood production variants from wood cultivation, including trees outside forests.

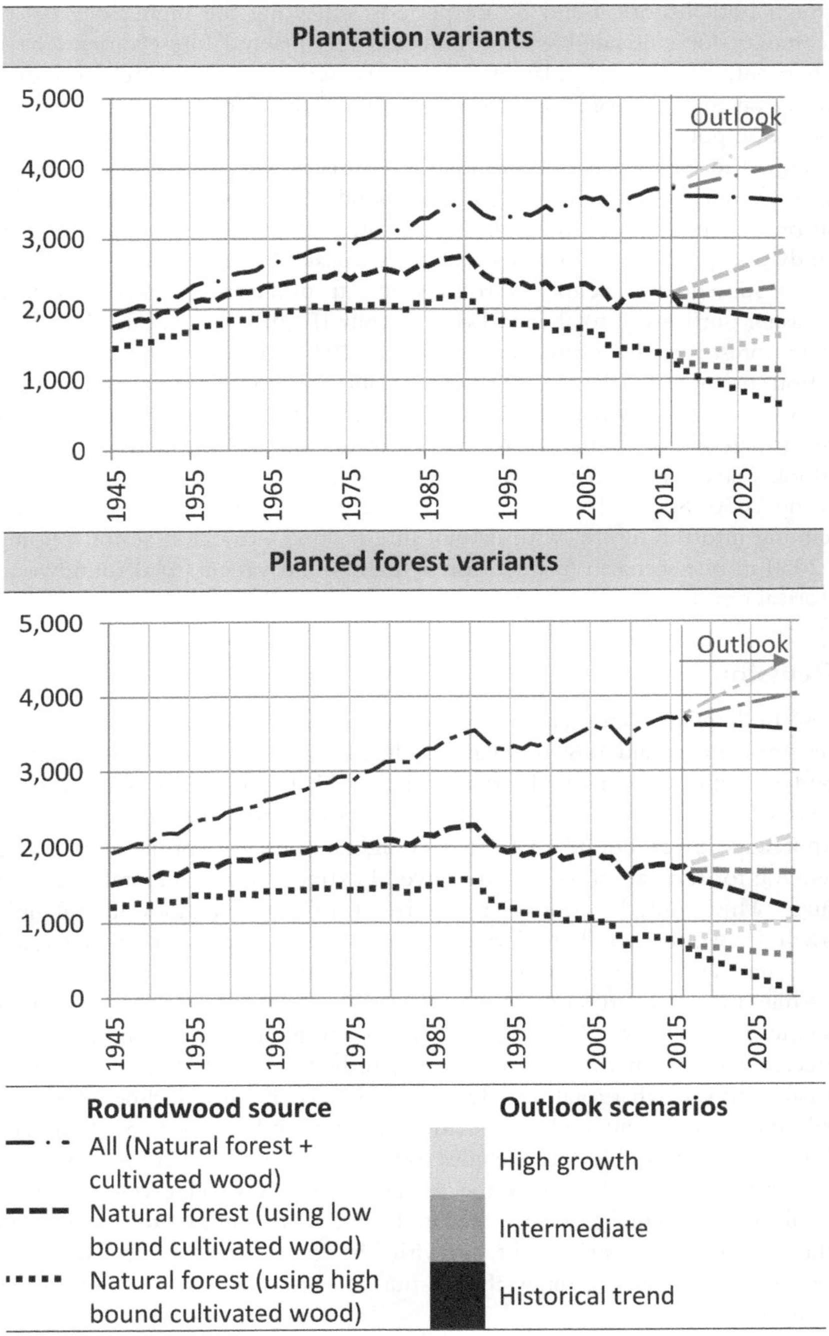

Figure 4.6 Estimates of global natural forest roundwood production 1945–2030.

through both historical and future periods, reflecting the higher degrees of uncertainty for cultivated wood estimates. The planted forest variants have higher outputs than the plantation variants, reflecting their inclusion of a broader range of the forest spectrum and, therefore, a larger area of forest and production potential.

Natural forest roundwood production estimates were derived by subtracting cultivated wood variants from total roundwood production figures. The historical natural forest roundwood estimates (1945 to 2016) indicate that roundwood production from natural forests peaked in either 1989 or 1990, with a subsequent decline as roundwood from cultivated trees replaced increasing portions of total roundwood supply (Figure 4.6). Of the 12 future natural forest roundwood scenarios (2017 to 2030), only one had future production passing the 1989/1990 peak (plantation low bound variant/total roundwood high growth, in 2029). The four high bound scenarios and one intermediate bound scenario have natural forest roundwood increasing in the outlook years at rates that would eventually pass the peak year at some point beyond 2030. Six of the 12 future scenarios have natural forest roundwood declining into the future, with natural forest wood extraction almost ceasing by 2030 in one scenario (planted forest high bound variant/total roundwood historical trend).

Discussion

These findings indicate that roundwood extraction from the world's natural forests peaked around 1989 and have declined since. This pattern of declining wood extraction from natural forests will most likely continue into the future. It would require both sustained high growth in total world roundwood output and lower bound cultivated roundwood supply projections to be operating for the pattern of declining wood extraction from natural forests to change. This would be contrary to the trends of recent decades of negligible growth in world roundwood output and significant growth in cultivated wood estate and outputs, and is considered unlikely.

While the Hubbert curve was developed for non-renewable resources (specifically oil) it can also be shown to represent any resource that is extracted faster than the resource is replenished (Bardi and Yaxley, 2005). This analysis finds that total roundwood production has not declined, but levelled, and is consistent with expectations for renewable resources (Hubbert, 1993). Nonetheless, this total includes roundwood from both cultivated wood and natural forests. This hides the pattern of roundwood extraction from natural forests, which, when separated out, does appear to have been extracted in a manner consistent with the theoretical pattern of non-renewable resources, or renewable resources extracted at rates above their replenishment level.

It is likely that a combination of factors, shifting both demand and supply, is creating a pattern of production from natural forests consistent with non-renewable or unsustainably extracted natural resources. Many of these have already been discussed. In summary, these factors include:

- Declining natural forest extent (FAO, 2010a; Gibbs *et al.*, 2010)
- Unsustainable harvesting (Bowles *et al.*, 1998; Shearman *et al.*, 2012)
- Increasing marginal costs of extracting natural wood (Sohngen *et al.*, 1999)
- Decreasing wood volumes from subsequent harvests in tropical natural forests (Putz *et al.*, 2012)
- Increasing supply and competitiveness of cultivated trees (Sedjo, 2001) driven by both market demands and government policies (Enters and Durst, 2004)
- Price feedbacks pushing wood-saving technologies and plantation investment (Hyde *et al.*, 1996)
- Pressure to restrict natural forest harvest through more constrained (and less profitable) forest practices (Pearce *et al.*, 2003; Rice *et al.*, 1997)
- Increased levels of forest reservation (FAO, 2010a), and bans on natural forest logging (as implemented in New Zealand, Thailand and Vietnam, for example).

As already noted, this transition to cultivated trees as a wood source could allow other natural forest values (especially biodiversity conservation and other ecosystem services) to be better managed while reducing conflict over natural forest resource allocation between wood production and other forest values (Ajani, 2011; Carle *et al.*, 2002; FAO, 2010a; Paquette and Messier, 2009; Sedjo and Botkin, 1997). Understanding the relative influence of the factors forcing the transition, noted above, is important to being able to quantify the extent to which there are forest conservation gains to made.[5] Reduced roundwood demand from natural forests can also provide opportunities for improvements in conservation and ecosystem service by providing space for previously logged natural forests to regenerate (Berry *et al.*, 2010; Gibson *et al.*, 2011). It can allow additional, precautionary prescriptions to be applied to wood extraction from remaining natural forests (Lindenmayer and Laurance, 2012) and the application of methods such as retention forestry and ecological forestry to improve non-timber value outcomes (Clark *et al.*, 2009; Franklin *et al.*, 2007; Gustafsson *et al.*, 2012; Lindenmayer *et al.*, 2006; Putz *et al.*, 2012). However, these 'light footprint' options can add to production costs, further reducing profitability and/or volumes of natural forest roundwood harvest. This could, in turn, have the effect of compounding the comparative cost advantages of cultivated wood sources, thus furthering the transition. Either way, whether more forest conservation or less wood extraction, wood from natural forests is likely to 'increasingly be a by-product of sustainable management for forest conservation' (Leslie, 2005, p. 14).

As well as presenting conservation opportunities, the transition of wood production to cultivated trees also has risks to conservation outcomes.

Cultivated wood establishment has significant biodiversity and ecosystem service implications. In particular, where establishing plantations involves conversion of natural forests, either directly as part of a process of forest degradation, or through forcing food production to displace natural forests elsewhere (for example, Bremer and Farley, 2010; Brockerhoff *et al.*, 2008; Gibbs *et al.*, 2010; Pawson *et al.*, 2013), there are high biodiversity and ecosystem service costs. This conversion is counter to the potential of cultivated trees to take pressure off natural forests as well as the potential ecosystem services gains to be had through planted forest afforestation (Bauhus *et al.*, 2010). At a national and regional level there is some data on establishing plantations through conversion of natural forests; for example, in Indonesia (Cossalter and Pye-Smith, 2003; Kartodihardjo and Supriono, 2000), China (Zhai *et al.*, 2012) and New Zealand (Hartley, 2002). However net global data is lacking. In its 2000 Global Forest Resource Assessment, the FAO found that between 1990 and 2000 the global plantation estate grew by 3.1 million hectares a year. Of this, it estimated 1.5 million hectares were the result of plantings on converted natural forests, with the balance through afforestation other land use classes (FAO, 2001, p. 9).[6] The 2010 Global Forest Resource Assessment (FAO, 2010a) did not repeat this estimate, although it reported that the rate of afforestation had increased in the ten year period from 2000 to 2010 while deforestation had slowed, indicating a possible decline in rates of conversion.

The above analysis clearly shows a global trend for natural forest wood extraction of increasing production until a peak around 1989, followed by a subsequent decline in production, suggestive of the Hubbert curve. This is reflective of the pattern shown in individual countries and represents a non-sustainable resource use pattern. The overall 'sustainability' of global wood production, then, is achieved by an increasing reliance on cultivated wood sources. A global peaking of natural forest wood extraction is highly significant. But its past recognition has likely been missed due to a lack of clear data. The modelling here helps fill that gap. The magnitude of the shift in wood sourcing from natural forests to cultivated sources raises significant questions for sustainable forest management, forest and biodiversity conservation, and ecosystem services delivery. The data presented here indicates a declining role for natural forests in meeting global wood demand. This in turn presents opportunities to reconsider how extensive natural forest ecosystems are managed, which is particularly important given rising societal demand for non-wood values from these forests.

Acknowledgements

From *Introduction* this chapter is a revised an updated version of an original paper. Reprinted by permission from Springer Nature: Biodiversity and Conservation (23, 5), *Global wood production from natural forests has peaked*, Russell D. Warman. Copyright © 2014.

Notes

1 'Roundwood' refers to the total wood removed from forests and from trees outside forests.
2 The terms 'extraction', 'production', 'supply', 'demand' and 'consumption' are all used here in relation to global roundwood. At the scale of the annual global data points used here it is assumed that, as the world is a closed system, the quantities of these will be the same; that is, 'extraction', 'production', 'supply', 'demand' and 'consumption' will generally match one another.
3 See also Jürgensen *et al.* (2014), published in the same year the work in this chapter was first published. Using a different method and focused on industrial roundwood only they also provide an account of recent and future global cultivated wood production. For 2012 they estimated 562 million m^3 of industrial roundwood coming from plantations. This was considered likely to be low as the total was known to exclude countries with plantations for which no data was available.
4 The three main concerns addressed in the review were: country survey response rates for fuelwood were low; per capita consumption rates had been considered to be static (though growing evidence contradicted this by pointing to a decline in per capita consumption in response to trends such as urbanisation and wealth increases); and prior estimations did not account for non-household use.
5 For example, if growth in plantation wood supply caused by government policy leads to a reduction of demand for natural forest logging then there are potential conservation gains. Conversely logging bans or new forest reserves created over forests for which there is no commercially accessible wood might achieve little real conservation benefit.
6 Plantations in the 2000 Global Forest Resource Assessment are a subset of the planted forest category. They represent only about a half of the extent of planted forests (Carle and Holmgren, 2008).

References

ABARE (Australian Bureau of Agricultural and Resource Economics) (1999). Global outlook for plantations. ABARE Research report. Canberra: ABARE.

Ajani, J. (2011). The global wood market, wood resource productivity and price trends: an examination with special attention to China. *Environmental Conservation*, 38, 53–63.

Arnold, M., Köhlin, G., Persson, R. and Shepherd, G. (2003). Fuelwood revisited: what has changed in the last decade? Bogor, Indonesia: CIFOR.

Bardi, U. and Yaxley, L. (2005). How general is the Hubbert curve? The case of fisheries. In: ASPO (ed.) *Proceedings of the 4th ASPO workshop. Lisbon: ASPO.*

Bauhus, J., Van Der Meer, P. and Kanninen, M. (2010). *Ecosystem goods and services from plantation forests*. Oxon: Earthscan.

Berry, N.J., Phillips, O.L., Lewis, S.L., Hill, J.K., Edwards, D.P., Tawatao, N.B., Ahmad, N., Magintan, D., Khen, C.V. and Maryati, M. (2010). The high value of logged tropical forests: lessons from northern Borneo. *Biodiversity and Conservation*, 19, 985–997.

Bowles, I. A., Rice, R.E., Mittermeier, R. and Da Fonseca, G. (1998). Logging and tropical forest conservation. *Science*, 280, 1899–1900.

Bremer, L.L. and Farley, K.A. (2010). Does plantation forestry restore biodiversity or create green deserts? A synthesis of the effects of land-use transitions on plant species richness. *Biodiversity and Conservation*, 19, 3893–3915.

Brockerhoff, E.G., Jactel, H., Parrotta, J.A., Quine, C.P. and Sayer, J. (2008). Plantation forests and biodiversity: oxymoron or opportunity? *Biodiversity and Conservation*, 17, 925–951.

Brown, C. (2000). The global outlook for future wood supply from forest plantations. Rome: Food and Agricultural Organisation of the United Nations.

Buongiorno, J., Zhu, S., Raunikar, R. and Prestemon, J.P. (2012). Outlook to 2060 for world forests and forest industries: a technical document supporting the Forest Service 2010 RPA assessment. Ashville, NC: United States Department of Agriculture.

Carle, J. and Holmgren, P. (2003). Definitions related to planted forests. UNFF intersessional expert meeting international steering group on "the role of planted forests in sustainable forest management. Maximising planted forests' contribution to SFM". Wellington.

Carle, J. and Holmgren, P. (2008). Wood from planted forests: a global outlook 2005–2030. *Forest Products Journal*, 58, 6–18.

Carle, J., Vuorinen, P. and Del Lungo, A. (2002). Status and trends in global forest plantation development. *Forest Products Journal*, 52, 12–23.

Clark, C., Poulsen, J., Malonga, R. and Elkan Jr, P. (2009). Logging concessions can extend the conservation estate for Central African tropical forests. *Conservation Biology*, 23, 1281–1293.

Cossalter, C. and Pye-Smith, C. (2003). Fast-wood forestry: myths and realities. Bogor, Indonesia: Center for International Forestry Research.

Del Lungo, A., Ball, J. and Carle, J. (2006). Global planted forests thematic study. Results and analysis. Rome: Food and Agricultural Organisation of the United Nations, FP38E.

Enters, T. and Durst, P. (2004). What does it take? The role of incentives in forest plantation development in Asia and the Pacific. Bangkok: Food and Agriculture Organization of the United Nations Regional Office for Asia and the Pacific.

Evans, J. (2009). *Planted forests: uses, impacts and sustainability*. Rome: CABI International and FAO.

FAO (1958). Yearbook of forest products statistics 1958. Rome: Food and Agriculture Organization of the United Nations.

FAO (1962). Yearbook of forest products statistics 1962. Rome: Food and Agriculture Organization of the United Nations.

FAO (2001). Global forest resources assessment 2000: main report. Rome: Food and Agriculture Organization of the United Nations.

FAO (2009). State of the world's forests. Rome: Food and Agriculture Organization of the United Nations.

FAO (2010a). Global forest resources assessment 2010: main report. Rome: Food and Agriculture Organization of the United Nations.

FAO (2010b). What woodfuels can do to mitigate climate change. FAO Forestry Paper. Rome.

FAOSTAT (2018). *Statistical databases*. Rome: Food and Agriculture Organization of the United Nations.

Franklin, J.F., Mitchell, R.J. and Palik, B. (2007). Natural disturbance and stand development principles for ecological forestry. Newtown Square, PA: US Department of Agriculture, Forest Service.

Gibbs, H., Ruesch, A., Achard, F., Clayton, M., Holmgren, P., Ramankutty, N. and Foley, J. (2010). Tropical forests were the primary sources of new agricultural land in the 1980s and 1990s. *Proceedings of the National Academy of Sciences*, 107, 16732–16737.

Gibson, L., Lee, T. M., Koh, L.P., Brook, B.W., Gardner, T.A., Barlow, J., Peres, C.A., Bradshaw, C.J., Laurance, W.F. and Lovejoy, T.E. (2011). Primary forests are irreplaceable for sustaining tropical biodiversity. *Nature*, 478, 378–381.

Gustafsson, L., Baker, S.C., Bauhus, J., Beese, W.J., Brodie, A., Kouki, J., Lindenmayer, D.B., Lohmus, A., Pastur, G.M., Messier, C., Neyland, M., Palik, B., Sverdrup-Thygeson, A., Volney, W.J.A., Wayne, A. and Franklin, J.F. (2012). Retention forestry to maintain multifunctional forests: a world perspective. *Bioscience*, 62, 633–645.

Hartley, M.J. (2002). Rationale and methods for conserving biodiversity in plantation forests. *Forest Ecology and Management*, 155, 81–95.

Heinimö, J. and Junginger, M. (2009). Production and trading of biomass for energy– an overview of the global status. *Biomass and Bioenergy*, 33, 1310–1320.

Hubbert, M.K. (1993). Exponential growth as a transient phenomenon in human history. In: Daly, H.E. and Townsend, K.N. (eds.) *Valuing the Earth: Economics, Ecology Ethics*. Cambridge, MA: MIT Press, 113–126.

Hyde, W.F., Amacher, G.S. and Magrath, W. (1996). Deforestation and forest land use: theory, evidence, and policy implications. *The World Bank Research Observer*, 11, 223–248.

Jürgensen, C., Kollert, W. and Lebedys, A. (2014). Assessment of industrial roundwood production from planted forests. FAO Planted Forests and Trees Working Paper Rome: Food and Agriculture Organization of the United Nations.

Kanowski, P. and Murray, H. (2008). Intensively managed planted forests: towards best practice. New Haven, CT: The Forests Dialogue.

Kartodihardjo, H. and Supriono, A. (2000). The impact of sectoral development on natural forest conversion and degradation: The case of timber and tree crop plantations in Indonesia. Bogor, Indonesia: Center for International Forestry Research.

Leslie, A. (2005). What will we want from the forests. *ITTO Tropical Forest Update*, 15, 14–16.

Lindenmayer, D., Franklin, J. and Fischer, J. (2006). General management principles and a checklist of strategies to guide forest biodiversity conservation. *Biological Conservation*, 131, 433–445.

Lindenmayer, D.B. and Laurance, W.F. (2012). A history of hubris – cautionary lessons in ecologically sustainable forest management. *Biological Conservation*, 151, 11–16.

Mead, D.J. (2005). Forests for energy and the role of planted trees. *Critical Reviews in Plant Sciences*, 24, 407–421.

Paquette, A. and Messier, C. (2009). The role of plantations in managing the world's forests in the Anthropocene. *Frontiers in Ecology and the Environment*, 8, 27–34.

Pawson, S., Brin, A., Brockerhoff, E., Lamb, D., Payn, T., Paquette, A. and Parrotta, J. (2013). Plantation forests, climate change and biodiversity. *Biodiversity and Conservation*, 22, 1–25.

Pearce, D., Putz, F.E. and Vanclay, J.K. (2003). Sustainable forestry in the tropics: panacea or folly? *Forest Ecology and Management*, 172, 229–247.

Putz, F.E., Zuidema, P.A., Synnott, T., Peña-Claros, M., Pinard, M.A., Sheil, D., Vanclay, J.K., Sist, P., Gourlet-Fleury, S. and Griscom, B. (2012). Sustaining conservation values in selectively logged tropical forests: the attained and the attainable. *Conservation Letters*, 5, 296–303.

Rice, R.E., Gullison, R.E. and Reid, J.W. (1997). Can sustainable management save tropical forests? *Scientific American*, 276, 44–49.

Sedjo, R. (2001). The economic contribution of biotechnology and forest plantations in global wood supply and forest conservation. In: Strauss, S.H. and Bradshaw, H.D. (eds.) *Proceedings of the First International Symposium on Ecological and Societal Aspects of Transgenic Plantations*. College of Forestry, Oregon State University.

Sedjo, R.A. and Botkin, D. (1997). Using forest plantations to spare natural forests. *Environment: Science and Policy for Sustainable Development*, 39, 14–30.

Shearman, P., Bryan, J. and Laurance, W.F. (2012). Are we approaching 'peak timber' in the tropics? *Biological Conservation*, 151, 17–21.

Smeets, E.M.W. and Faaij, A.P.C. (2007). Bioenergy potentials from forestry in 2050. *Climatic Change*, 81, 353–390.

Sohngen, B., Mendelsohn, R. and Sedjo, R. (1999). Forest management, conservation, and global timber markets. *American Journal of Agricultural Economics*, 81, 1–13.

Solberg, B., Brooks, D., Pajuoja, H., Peck, T. and Wardle, P. (1996). An overview of factors affecting the long-term trends of non-industrial and industrial wood supply and demand. Long-term trends and prospects in world supply and demand for wood and implications for sustainable forest management. Finland: European Forest Institute.

Varmola, M., Gautier, D., Lee, D., Montagnini, F., Saramäki, J., Mery, G., Alfaro, R., Kanninen, M. and Lobovikov, M. (2005). Diversifying functions of planted forests. Forests in the global balance – changing paradigms. *IUFRO World Series*, 17.

Weiner, R.U. and Victor, D.G. (2000). Industrial roundwood demand projections to 2050: a brief review of the literature. New York: Council of Foreign Relations.

White, A., Sun, X., Canby, K., Xu, J., Barr, C., Katsigris, E., Bull, G., Cossalter, C. and Nilsson, S. (2006). China and the global market for forest products. Transforming Trade to Benefit Forests and Livelihoods. Washington DC: Forest Trends.

Whiteman, A., Broadhead, J. and Bahdon, J. (2002). The revision of woodfuel estimates in FAOSTAT. *Unasylva*, 53, 41–45.

Zhai, D.-L., Xu, J.-C., Dai, Z.-C., Cannon, C.H. and Grumbine, R. (2012). Increasing tree cover while losing diverse natural forests in tropical Hainan, China. *Regional Environmental Change*, 1–11.

Zomer, R.J., Trabucco, A., Coe, R. and Place, F. (2009). Trees on farm: analysis of global extent and geographical patterns of agroforestry. ICRAF Working Paper-World Agroforestry Centre.

5 A natural turn

Land use change, leakage and forest conservation

Changes in wood production, forest use and forest extent are often studied within the natural sciences. There is a compelling logic to consider these phenomena as grounded in the realm of the natural. The growth of trees and forests, their location in landscapes, relations with climate and soils and enmeshment within ecologies lend themselves to dissection by the natural sciences. Likewise, classic economic thought brings a similar reductionist spirit to the study of wood and forest social-ecological systems. The hard-to-measure power and word plays of human actors can be dealt with by assuming that these actors are rational beings seeking to optimise their own well-being, constrained only by measurable natural phenomena. This is the epistemological basis of the following chapter.

While this chapter considers the implications of policy, it does so through analysing areas of forest and land tenure and volumes of wood production and trade in Australia. It looks at how changes in these might illuminate key policy considerations in relation to the transition of stewardship forestry into post-forestry wood cultivation and forest ecosystem management. In particular, it looks at how industrial plantation policy and forest conservation policy interact. This chapter does this through an analysis of potential leakage of wood harvest impacts relating to policy change reflecting the transition. Leakage research is an important analytical tool for understanding complex social-ecological systems. Assessing leakage seeks to understand (and quantify) how systems might not deliver the results intended, particularly through unanticipated effects on interconnected systems.

Introduction

Over recent decades significant areas of Australia's publicly owned natural forest have been reallocated from production forest to conservation forest. During the same period, a range of policies supported the establishment of plantations. This case study analyses whether the intended conservation outcomes of Australian forest policy have been undermined by conservation loss in other natural forests due to shifting wood demand. It is intuitively compelling to design policies that seek to conserve forest ecosystems by creating areas of natural forest that are reserved or protected from wood production in order to maintain or increase their non-wood values.[1] Still, wood harvesting is a mobile economic activity and attempts to conserve forests through restrictions on production in one area can shift that activity elsewhere. This can negate the conservation benefits of forest protection achieved by the original restriction on production. This phenomenon of inadvertently shifting impacts elsewhere is a policy problem referred to variously as 'leakage', 'slippage' or 'displacement'. Understanding and managing leakage is critical to the success of policies designed to conserve natural forest.[2]

The area of the world's forests legally conserved for non-wood ecosystem services is increasing. Globally, the area of forests in protected areas grew from 441 million hectares in 1990 to 651 million hectares in 2015 (FAO, 2015). In Australia the reported area of natural forest in formal conservation reserves grew from 17.6 million hectares in 1997 (National Forest Inventory, 1998) to 21.5 million hectares in 2013 (National Forest Inventory, 2013). The net area of public natural forest allocated for wood production in Australia declined by 45 per cent over the period 1996–1997 to 2011–2012 from ten to 5.5 million hectares (National Forest Inventory, 2013, p. 125).

Wood production from Australia's natural forest declined 56 per cent, from 9.6 million m^3 in 1996–1997 to just under 4.3 million m^3 in 2016–2017. Over the same period wood production from plantations nearly tripled, from 10.8 million m^3 to 29.1 million m^3 due to policies implemented to increase the area of plantation in Australia. There were two phases of plantation expansion: a federal loan scheme to state governments for softwood plantations from 1965 until the end of the 1980s, which was then followed by the National Forest Policy of 1995 which included a vision for a threefold increase in plantation area by 2020. The 1995 policy was supported by a favourable federal government tax treatment, Managed Investment Schemes (MIS), which supported establishing hardwood plantations (Ferguson, 2014). These policy drivers are consistent with the global trends of wood production shifting to more intensive planted sources driven by the factors already noted.

The 1995 revision of Australia's National Forest Policy Statement led to the implementation of a series of regionally-based forest policy initiatives known as Regional Forest Agreements (Dargavel, 1998; Kirkpatrick, 1998; Lane, 1999). While the Regional Forest Agreements were focused on a small area of Australia located in south-east and south-west (Clancy and Howell, 2013, p. 5), they covered almost all of Australia's commercial natural forest wood-producing regions. The overarching policy objective of the Regional Forest Agreements was to resolve conflict surrounding the use of natural forest that had been intensifying since the 1970s (Clancy and Howell, 2013; Lane, 1999). The Regional Forest Agreements sought to optimally allocate public natural forest to conservation or wood production. One outcome of the policy was an increase in the area of public natural forest allocated to conservation reserves based on detailed technocratic analysis of their economic, social and environmental values.

It has been asked subsequently if the direct gains in conservation have been offset by displacing wood production to other natural forests in Australia or around the world (Institute of Foresters of Australia, 2011; Whittle *et al.*, 2012). The risk of leakage is a globally recognised phenomenon (Lambin and Meyfroidt, 2011; Pfaff and Walker, 2010), and has been used to caution against unilateral conservation action by governments (Gan and McCarl, 2007). This work seeks empirical evidence for these concerns. Australia's Regional Forest Agreements provide an ideal opportunity to analyse the phenomenon of leakage, and to examine the conditions under which public policy decisions to conserve natural forest are likely to be effective.

Background

Over the last 15 years there has been considerable literature on leakage in relation to forest carbon stores, avoided deforestation and avoided log extraction impacts on greenhouse gas emissions (for example, see the literature reviews of Atmadja and Verchot, 2012; Henders and Ostwald, 2012). Concern over leakage and a need to measure greenhouse gas emission in ways that are consistent and efficient has contributed to life cycle assessment and related international standards (Guinée, 2002). There is specific literature on protecting forests from wood extraction and subsequent leakage or displacement of log extraction effort to other forests (Chomitz *et al.*, 2007; Gan and McCarl, 2007; Meyfroidt and Lambin, 2009). Measuring and incorporating leakage into forest conservation projects is a particularly challenging technical issue (Aukland *et al.*, 2003; Rapidel *et al.*, 2011) and ongoing research is required to improve understanding (Atmadja and Verchot, 2012).

Quantifying leakage is an attempt to account for impacts that are unintended and potentially external to the sphere of influence/concern of decision makers (Figure 5.1). The use of the term 'leakage' makes this accidental nature of the phenomenon clear. Leakage can occur as a consequence of bounded rationality (Simon, 1972) or as a form of externality to the policy maker's jurisdiction or sphere of interest. Leakage assessment aims to quantify the impacts of an intervention beyond those explicitly understood and accounted for. It reflects recognition that the sphere of influence (or concern) for most policy interventions is a bounded or limited subset of the overall social and ecological systems involved (Wunder, 2008). Limited knowledge of social and ecological systems or limited concern with externalities means that any planned intervention can result in positive or negative leakage – unintended effects that can either reinforce or negate intended outcomes. Because of this, assessments of leakage provide essential feedback for monitoring and reviewing policy design.

Leakage assessment involves comparing the net outcomes following an intervention with a scenario in which there is no intervention – the counterfactual. Assessments can be conducted *ex ante* to predict leakage or *ex post* to review leakage following a policy intervention. The former approach requires

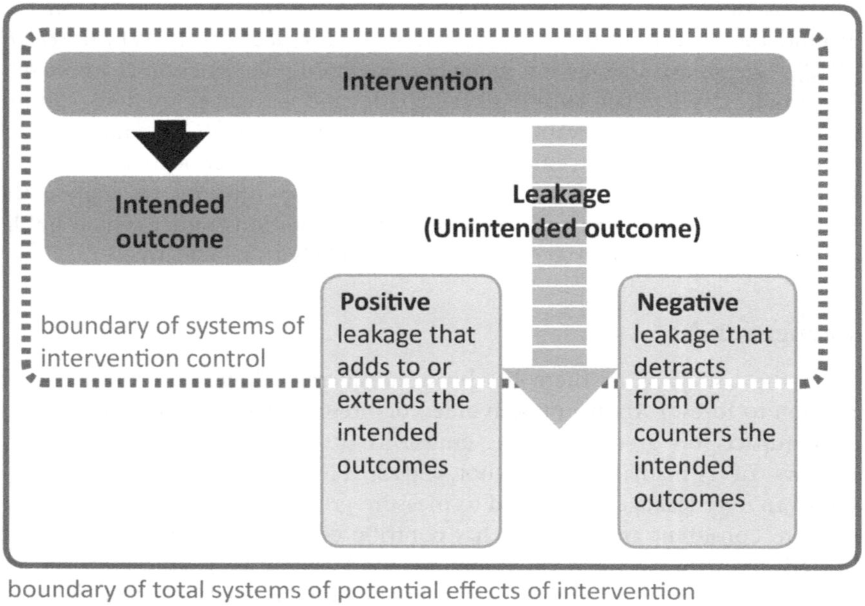

Figure 5.1 Basic model of leakage and its relationship to intended outcomes.

modelling of both the counterfactual and intervention scenarios, while *ex post* assessment can use data from the intervention scenario and compare this to a counterfactual.

A proliferation of terminologies and typologies accompany the growing field of leakage assessment, with potential for confusion (Aukland *et al.*, 2003). A distinction has been made between weak leakage – general displacement of demand – and strong leakage – the direct unintended consequences of a policy intervention (Henders and Ostwald, 2014). The former tends to focus on shifts in demand without attributing causality to specific drivers. Weak leakage can be a useful concept in some cases such as monitoring the net outcomes of atmospheric CO_2 balance. Strong leakage can be useful in seeking to understand the effects of a specific change in policy.

Another commonly used distinction for different types of leakage is *direct* and *indirect*. Direct leakage occurs when the targeted activity (for example, wood production) is shifted elsewhere. Indirect leakage occurs when the intervention causes other types of activity that have impacts on intended outcomes. Leakage that occurs within the sphere of influence of an agent is often referred to as local leakage. Leakage outside the same sphere is referred to as distant leakage – often at risk of being discounted by local agents. Other typologies reflect the alternative means of triggering changes in agent behaviour. Changes triggered directly by an intervention are referred to as primary leakage, while those triggered indirectly by market signals are referred to as secondary leakage. Figure 5.2 summarises a typology drawing on recent reviews of leakage (Atmadja and Verchot, 2012; Aukland *et al.*, 2003; Henders and Ostwald, 2012).

Past assessments

While leakage is well appreciated as an issue of concern in the literature on forest policy, there have been relatively few attempts to estimate leakage arising from policy decisions to conserve natural forest. Two related approaches have emerged – economic modelling of future scenarios and analyses of production and trade data.

Murray *et al.* (2004) used econometric analysis to estimate the effects of leakage between two regions when an area of forest in one is conserved – assuming that both are of equal conservation value. When the wood from both regions are perfect substitutes, increased conservation in one region increases the price of wood in both regions leading to reduced consumption and increased production in the region without forest conservation – that is, you would expect some, but not all, of the wood production to leak. In this model, the more imperfect the wood in the regions are as substitutes, the more muted is the leakage effect (Murray *et al.*, 2004, p. 115).

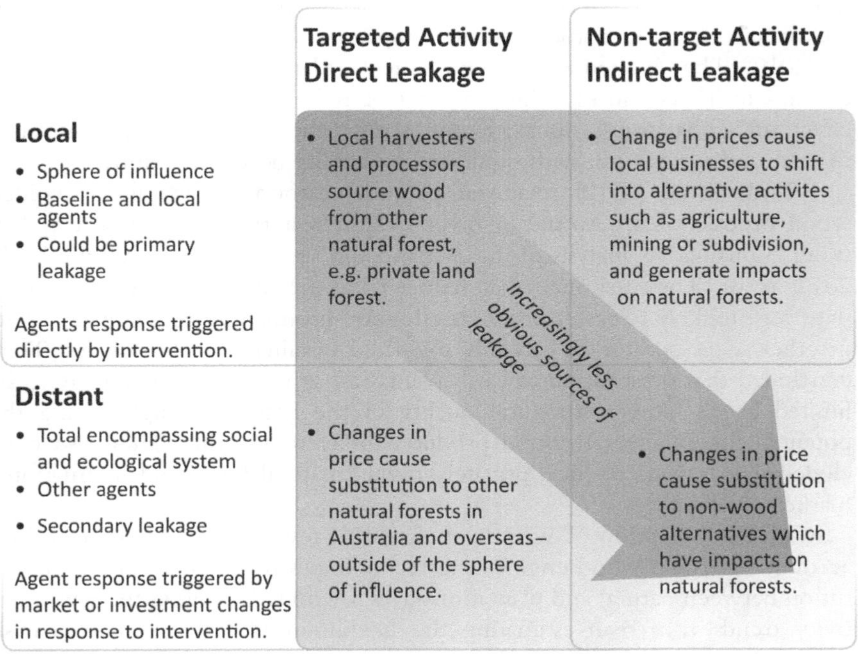

Figure 5.2 Matrix of main distinctions of leakage sources and types.

Gan and McCarl (2007) used a computable general equilibrium model to quantify leakage from forest conservation between countries. They determined that forest conservation efforts in Australia and New Zealand (treated as a single region) would have a leakage rate of 89 per cent to the rest of the world, with 70 per cent to tropical forests in developing countries. Leakage levels for the rest of the world ranged from 42 per cent for Canada to 95 per cent for Russia. They found negligible leakage into markets for non-wood alternatives, such as other fibre sources, metals and plastics. They also found that coordinated efforts between countries to conserve forests have the potential to reduce leakage. A key limitation in this analysis was that it did not distinguish between plantation and natural forest wood sources.

Smaller estimates of leakage in the literature include that of Murray *et al.* (2004. p. 122), who reported that leakage of carbon from forest conservation projects could be 'somewhat larger than the energy sector estimates (previously cited as roughly 5 to 20 per cent)'. Sohngen *et al.* (1999) estimated that conserving 5 per cent or 10 per cent of North American and European forests would result in leakage of about one hectare of additional inaccessible land worldwide for every 20 hectares set aside – a leakage rate of about 5 per cent.

The analyses above are based on shifts in demand and supply resulting in substitution in the source of wood production which has then been equated to leakage of conservation values. This provides a modelling framework for predicting future leakage only in cases where it can be assumed that production levels reflect conservation values. Leakage effects can also be assessed retrospectively using a stock and flow approach, such as that of Meyfroidt *et al.* (2010). They looked at the displacement of land use resulting from reforestation in 12 countries. They found that reforestation had a net positive effect on global forest stocks despite significant leakage through displaced forestry and agriculture, with significant variability between countries.

Whittle *et al.* (2012) reviewed potential carbon leakage from avoided wood production in Australian natural forest resulting from climate policy using 'a qualitative analysis of the demand and supply factors' (Whittle *et al.*, 2012, p. 6). They reasoned that reduced log production in one Australian state was unlikely to result in increased wood production from similar forests in other states because of carefully regulated sustainable yield policies. They concluded that substitution towards plantation-grown wood was likely to be limited by the lower structural quality of the latter, although noting the potential for engineered wood products to overcome this. They also concluded that leakage to local privately-owned natural forest and international markets was likely.

With the exception of Whittle *et al.* (2012) few of these studies take into account emerging wood engineering technologies that enable greater substitution between natural and plantation-grown wood. Excluding these productivity trends risks over-estimating the likelihood of leakage. The analysis provided below is *ex post*, relying on observed data which automatically captures trends in technical efficiency that affect product substitution and leakage. The significant limitations and varied results in what has been a small and incomplete body of model-based analyses show there is value in conducting the following empirical case study analysis.

Methods

This chapter analyses wood production to provide a prima facie assessment of leakage resulting from the conservation of public natural forest. It starts with an assessment of changes in land use policy and their impacts on sustainable yield and actual harvest. It then reviews patterns in consumption in Australia before evaluating changes in each of the three main sources of natural forest to which log harvest might have leaked as a result of Regional Forest Agreement land tenure changes – natural forest on private land, in other states and internationally.

Policy decisions to conserve natural forest can be undertaken to meet multiple objectives such as conserving biodiversity, storing carbon, improving water quality and providing recreation areas. Attributing leakage to these multiple interacting values is difficult. This assessment will focus on direct leakage of the targeted activity – log removals – and potential shifts in log removals from newly conserved areas of natural forest to natural forest in other locations or jurisdictions. Log removals are used as a causal indicator of impact on natural forest conservation values.

Data on wood production was available for 1996–1997 to 2016–2017,[3] a period that corresponds to the implementation of the Regional Forest Agreements and their progressive reallocation of natural forest from production to conservation. Data is reported in financial years – from 1 July to 30 June – consistent with the reporting years used by Australian Bureau of Agricultural and Resource Economics and Sciences (ABARES), the main source of data for this analysis.

Changes in the area of public natural forest available for wood production were obtained from *Australia's State of the Forests Reports*, where the net area of forest available for wood production was estimated from the area of production forest minus areas set aside from wood production by informal reserves and other administrative restrictions (National Forest Inventory, 2013). Estimates of sustainable yield indicate the amount of wood that can be produced on an ongoing basis. Sustainable yield at the beginning of the Regional Forest Agreement period provides an estimate of the volume of wood that could have been harvested from public natural forest without a change in land tenure. Prior to the Regional Forest Agreement policy, natural forest roundwood extraction in Australia had been stable for at least 20 years (Love *et al.*, 1999).

Australian wood production and use data is disaggregated between hardwood (or broad leaved) and softwood (or coniferous). Hardwood is also disaggregated by source, from plantations or natural forests. Softwood in Australia is primarily from plantations of exotic pine (*Pinus* sp.) species, with some limited production from indigenous hoop pine (*Auracaria* sp.) and cypress pine (*Callitris* sp.). Cypress log production (*c.*2.9 per cent of total natural forest log production) was added to sawlog production from natural hardwood forests and subtracted from softwood production in the two states where it is produced (New South Wales and Queensland). Data for production, consumption, imports and exports for wood from natural forest and plantations were obtained from ABARES publications (ABARES 2018a; 2018b; 2018c) and supplemented by additional data and disaggregations supplied directly from ABARES (from herein all data provided from these four ABARES sources will be cited as *ABARES data*). Production of plantation logs was calculated by summing the production from hardwood and softwood plantations.

ABARES does not include volume information for secondary wood products (wooden furniture, printed paper products [such as books and magazines], paper manufactures and prefabricated buildings), but Australian dollar value data gave some indication of trends. Over the period of this assessment combined net trade in printed paper products and paper manufactures stayed relatively consistent in constant dollar terms (net imports of printed paper products declined by 29 per cent and net imports of paper manufactures grew by 45 per cent), while net trade of prefabricated wooden buildings was negligible in most years except a for a spike in 2011–2012 and 2012–2013. Other than in these years they represented small volumes. Wooden furniture net imports, however, grew significantly across this period (478 per cent), representing a significant potential source of leakage. For this category, the data used to calculate the volume of wood imported as furniture was obtained from the United Nations Comtrade database (United Nations, 2018), using conversion factors for furniture to whole logs from Contreras-Hermosilla *et al.* (2007) and reverse extrapolation to calculate volumes for the first four years of the period for which volume data for wooden furniture was unavailable.

This analysis focuses on the two main log types derived from Australian natural forest – sawlogs and pulplogs. Sawlogs are used to produce sawn timber and veneers (for plywood), while pulplogs are used mainly to produce reconstituted wood panels (mostly medium density fibreboard and particleboard) and diverse paper products.

To assess leakage it was necessary to make comparisons between what happened and a counterfactual scenario – what would have happened without the increase in the area of public natural forest allocated to conservation. The assumed counterfactual is as follows. It is expected that the potential log production would have remained constant from public natural forests based on sustainable yield at the time, and ongoing government commitment to these sustainable yields. This assumption is supported by already noted evidence of stable production in previous decades. Therefore it is assumed that sustainable yield from public natural forest in Australia would have remained at 2.96 million m^3 for sawlogs (see Figure 5.3), and 3.56 million m^3 for pulplogs.[4] The maximum theoretical wood volume that could be leaked to other sources is the difference between this sustainable yield at the beginning and end of the period. Annualised sustainable yield figures for Australian public natural forests are not available. Here the latest set of figures available are used. These are for the period of 2006–2007 to 2010–2011 (1.68 million m^3). It is likely that subsequent revisions of sustainable yield have continued the downward trend noted in Figure 5.3, however the latest published figures are used as they will give a more conservative estimation of leakage. Using these figures gives a maximum amount of wood that could leak of 1.28 million m^3 of sawlog, 1.56 million m^3 of pulplog and 2.84 million m^3 of total logs. It is assumed wood production from plantations in a counterfactual would have increased at the same rate observed in the actual. It is assumed that all other trends in demand for wood product would have continued as per the actual in the absence of the Regional Forest Agreement suite of policies. It is assumed that technological improvements would have been the same in both the actual and the counterfactual. It is also assumed that the Regional Forest Agreement policy changes had no influence on international sawlog or pulplog production or demand.[5]

Results and analysis

The sustainable yield and the production of sawlogs from public natural forest in Australia declined by about 43 per cent over the Regional Forest Agreement period (National Forest Inventory, 2013) (Figure 5.3), which is in line with the decrease in area of public natural forest available for production (45 per cent). A critical first observation is that production of sawlogs has been consistently below the sustainable yield (an average of 87.1 per cent of

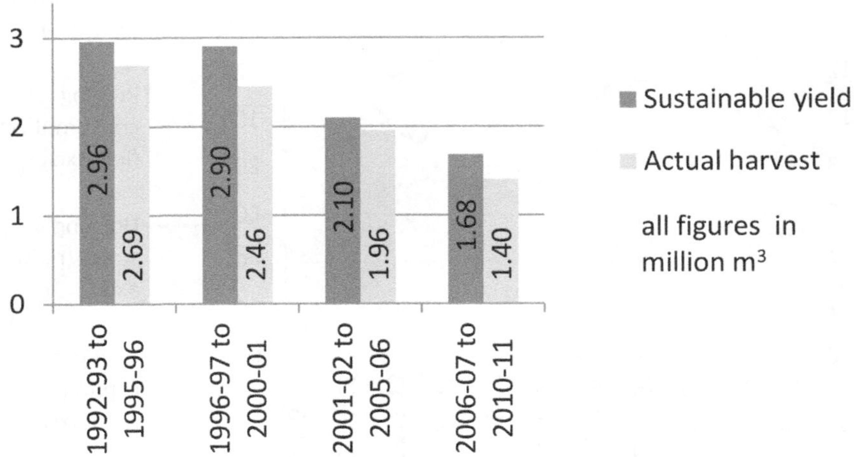

Figure 5.3 Sustainable versus actual yield of sawlogs from Australia's public natural forest.

Source: ABARES data; National Forest Inventory, 2013.

sustainable yield throughout the Regional Forest Agreement period). This indicates that the decline in public natural forest sawlog production was not caused by decreased supply resulting from the Regional Forest Agreement policy but rather from a reduction in demand for natural forest sawlog. This in itself suggests there was no unmet demand from public natural forests as a result of the Regional Forest Agreement policy of increased land protection. Rather, this indicates the land area available for harvest has simply reduced in line with decreasing demand. While other factors might have caused production to be so consistently below sustainable yield, government policy settings throughout this period remained set to maximise sustainable production. This would support a conclusion that there was no unmet demand to leak as a result of the policy change. Despite this, detailed assessment has been undertaken to consider other possibilities that might indicate negative leakage (including smaller or temporary indications of leakage).

ABARES produces estimates of whole log equivalents of consumption for sawlogs (based on sawn wood and veneer consumption) and pulplogs (based on wood panel and paper product consumption) (Figure 5.4). Australian consumption of sawlog equivalents increased from 11.5 million m^3 in 1996–1997 to 14.6 million m^3 in 2003–2004, and then reduced to 12.9 million m^3 in 2016–2017. Sawlog consumption in Australia is correlated to housing starts (Burke and Townsend, 2011), and the correlation is shown in Figure 5.4. Pulplog consumption has trended slightly upwards over the period but at a

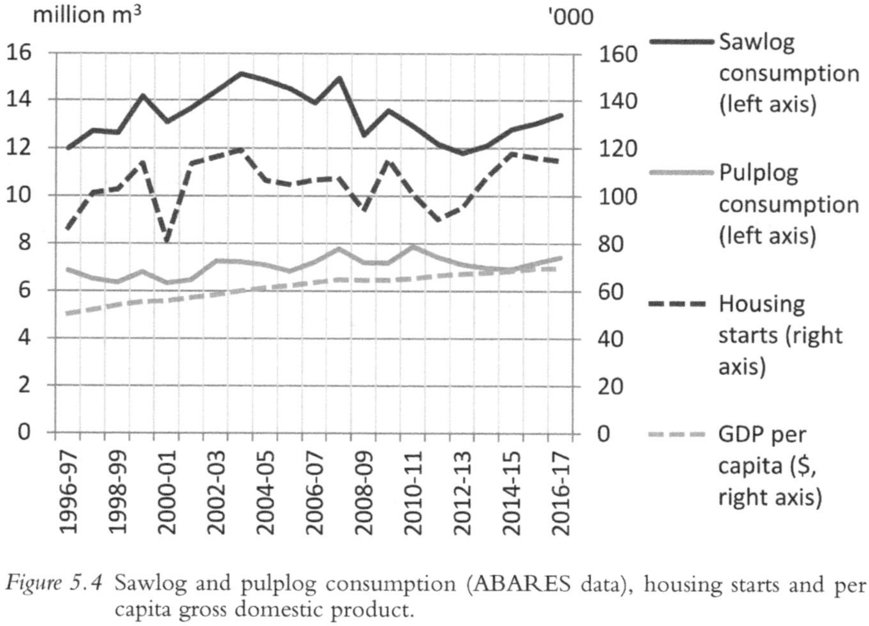

Figure 5.4 Sawlog and pulplog consumption (ABARES data), housing starts and per capita gross domestic product.

Sources: Australian Bureau of Statistics, 2018a, 2018b.

slower rate than per capita income. This is consistent with global patterns of paper consumption in developed economies (Andres *et al.*, 2014). Australian per capita consumption of whole log equivalents of wood products continued to decline through the period.[6] This is consistent with long term trends in wood consumption, including changes in paper consumption, recycling and wood use efficiency noted already. Overall, Australian wood consumption does not display any significant change beyond pre-existing trends.

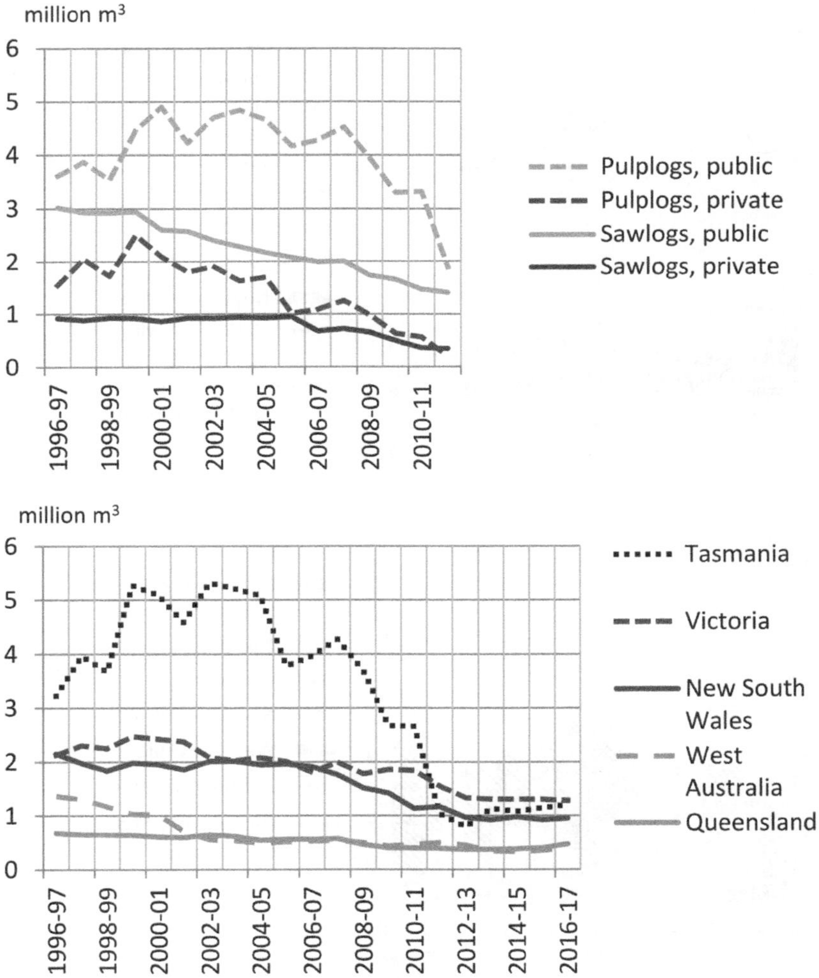

Figure 5.5 Top – Sawlog and pulplog production from natural forest on public and private land in Australia 1996–97 to 2010–11. Bottom – Log production from natural forest for the main producing states of Australia 1996–1997 to 2016–2017.

Source: ABARES data.

Leakage to private land natural forest

All states in Australia where natural forest is used for wood production also have similar forest types on privately owned land.[7] These are potential sources of leakage. However, sawlog and pulplog production from natural forest on privately owned land in Australia has followed similar trends to production from public natural forest (Figure 5.5), contrary to obvious opposite trends which would indicate leakage. Overall, the proportion of Australian natural forest log

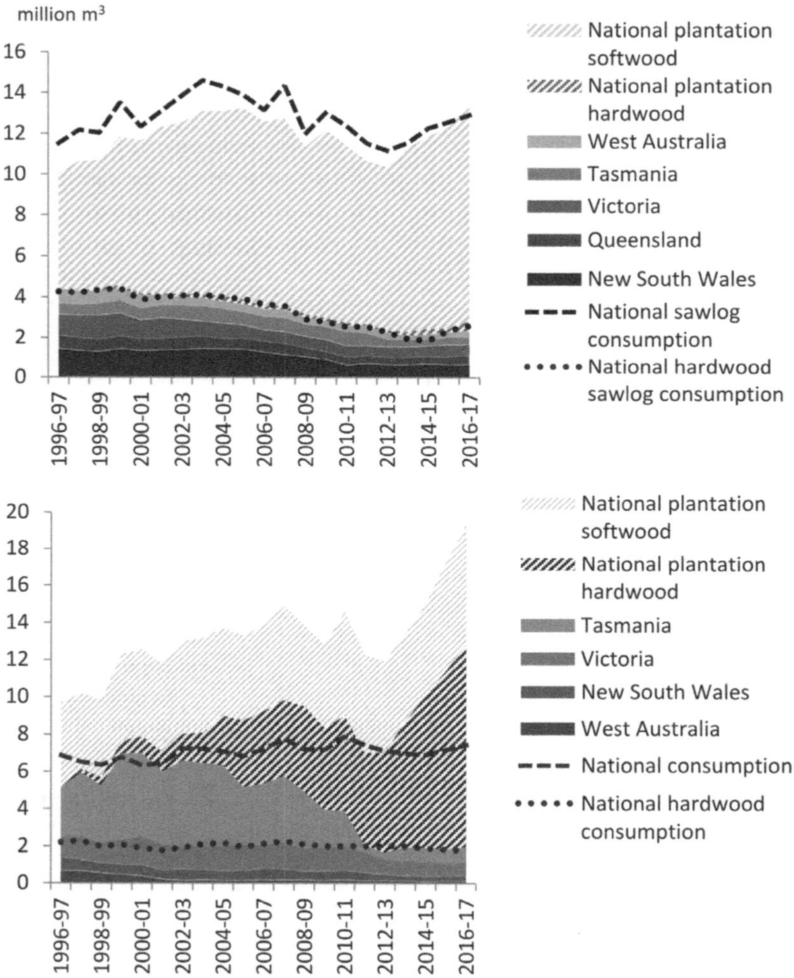

Figure 5.6 Sawlog (top) and pulplog (below) production from natural forest in each Australian state and national production from plantations; total national consumption of logs and national hardwood log consumption.

Source: ABARES data.

production from public natural forest increased from 73 per cent in 1996–1997 to 85 per cent in 2011–2012 (ABARES data). Falling production from privately owned natural forest over an extended period when conservation of public natural forest has increased provides prima facie evidence against direct leakage from public to private natural forest. State-by-state analysis of private land sawlog harvest was provided in the National Forest Inventory (2013)[8] with average levels of harvest for each of the first three five year blocks of the period under review here (1996–1997 to 2000–2001; 2001–2002 to 2005–2006; and 2006–2007 to 2010–2011). All states except Western Australia saw a decline between the first and last period. The increase in Western Australia represents a theoretical leakage of 1.8 per cent of national natural forest sawlogs.[9] New South Wales and Western Australia both had increases in the middle period. This also represents possible temporary leakage.

It is also possible that if overall demand for natural forest sawlogs declined over the period due to plantation substitution, then the rate of reduction in supply from private natural forest might have been faster in the counterfactual if public natural forests grew their share of the reducing demand in this period – thus leakage. Nonetheless, as noted above, the rate of sawlog supply from public natural forest has been consistently below sustainable yield at an aggregate national level, making it unlikely that there was any unmet demand from reduced sustainable yield to be leaked.

Leakage between states

Log production from natural forest declined in all Australian states over the period 1996–1997 to 2016–2017: between 71 per cent in Western Australia and 30 per cent in Victoria (Figure 5.6). Production in Tasmania was unique in that it grew significantly early in the period, mostly due to an increase in pulplog production (Figure 5.6) and is evidence of possible leakage from the conservation of public natural forest in other Australian states. Even so, the scale of this 'bubble' of production was much larger than the maximum potential leakage from other states at the time, indicating drivers other than the change in area of public natural forest available for harvest in other states. In particular, this increased pulplog production in Tasmania coincided with a period of increased conversion of natural forest to plantations (Forest Practices Authority, 2014) generating extra volumes of wood and a growth in demand for natural forest hardwood chips by Japan, before a collapse marked by both a decline in overall Japanese hardwood woodchip demand and a shift in preference to plantation woodchip (Macintosh, 2013).

Natural forest sawlog production in all states fell over the period 1996–1997 to 2016–2017, indicating no net displacement of production between states (Figure 5.6). During the same period production of plantation sawlogs grew considerably (Figure 5.6), mostly from softwood plantations. Consumption of hardwood sawlogs declined with production of hardwood sawlogs while consumption shifted to softwood logs in line with increasing production.

International leakage

During the period of interest, Australia was mostly a net importer of sawlog equivalents and a net exporter of pulplog equivalents (Figure 5.7). Net imports of sawlog equivalents fell by 1.88 million m^3 between 1996–1997 and 2016–2017 and by the end of the period Australia was a net exporter of sawlog equivalents. This corresponded with a period of significant substitution towards plantation grown sawlogs. Production of hardwood sawlogs and consumption have closely mirrored one another with consumption reducing in line with production indicating no or little leakage.

Over the same period, net exports of pulplog equivalents increased from 2.82 to 11.49 million m^3 (Figure 5.7). Neither sawlog nor pulplog net trade patterns indicate negative leakage to international forests. Indeed, increased export of plantation pulplogs and decreased imports of sawlogs to Australia may have reduced demand for logs from natural forest in other countries, resulting in net positive leakage (of course, this would be attributed to Australian plantation establishment policies rather than Regional Forest Agreement conservation policy).

The increase in imports of wooden furniture, a secondary wood product, over the Regional Forest Agreement period represents a potential source of leakage. Whole log equivalents for net furniture trade is shown in Figure 5.7

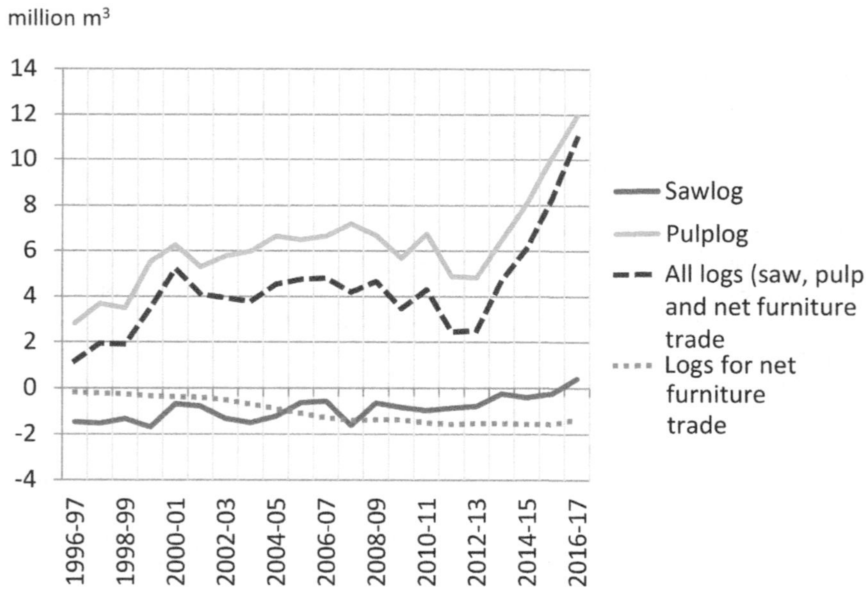

Figure 5.7 Australian net trade of sawlog equivalents, pulplog equivalents, log equivalents for net furniture trade and total net log equivalents.

Sources: ABARES data; Contreras-Hermosilla *et al.*, 2007, United Nations, 2018.

and increased from –0.19 million m^3 in 1996–1997 to –1.37 million m^3 in 2013–2014. However, this was more than offset by positive growth in net trade for all other log sources. Including wooden furniture trade, Australia's net trade of all log equivalents increased between 1996–1997 and 2016–2017 by 9.82 million m^3, indicating no international leakage.

It should also be noted that Australia's furniture production at the start of the period involved only a portion of its wood coming from natural forest, with the balance made up of plantation pine, particle board and medium density fibreboard. It was estimated that 5.5 per cent of natural forest sawlogs harvested in Australia in 1998 were used for furniture production (Jaakko Pöyry Consulting, 1999).[10] This would amount to 162,250 m^3 of the sustainable yield of public natural forest sawlog at the start of the Regional Forest Agreement period. Assuming this natural forest sawlog supply for furniture was produced entirely from public natural forest at the start of the period, this gives a maximum leakage of 5.8 per cent if all of this wood demand was shifted outside of Australia. However, a number of factors negate this possibility. First, over the Regional Forest Agreement period whole log equivalents of Australian net exports of hardwood sawn wood increased by 157,600 m^3. That is, net exports of hardwood sawn wood over the same period was almost identical to the maximum possible leakage of natural forest hardwood logs for furniture production. This is suggestive of shifts towards the manufacturing of furniture abroad (changes independent of the Regional Forest Agreement policy rather than shifts of wood sourcing which would constitute leakage). Also contradicting the possibility of leakage, there is evidence of an increase, not a decrease, in the use of natural forest hardwoods in Australia in furniture and other high value applications over the period (URS, 2007).[11] Finally, the data does not capture the export of whole logs and sawn wood from Australia to offshore furniture manufacturing (Markets for Change, 2011) which would reduce the stated net log trade impact of furniture and not constitute leakage.

Discussion

The results indicate that there has been no significant leakage of wood production to other natural forest – either locally or internationally – from the conservation of Australia's natural forest under the Regional Forest Agreements. The main reason for this is substitution towards domestic plantation wood supplies. It is also possible that, as Australia increased net exports of whole log equivalents, positive leakage from plantation establishment may have occurred as wood from Australian plantations displaced wood that would otherwise have been harvested from natural forest internationally. The overall finding of minimal wood production leakage is at odds with a number of past, mainly model-based or first -principle assessments that have forecast relatively high levels of leakage. The analysis undertaken here suggests that overestimates of leakage can result from underestimates of various substitution effects. In the Australian case, substitution with emerging plantation wood supplies has been critical in negating leakage. This highlights the importance of distinguishing between natural and planted wood sources when estimating potential leakage of forest conservation impacts.

The Australian case shows how shifts in processes along the supply chain have interacted with the reduced supply from natural forest in ways other than displacement of wood sourcing to other natural forest. A likely factor to consider in leakage assessment is the extent to which a shift in relative prices leads to improvements in productivity rather than simply displacing the activity of concern (Schwarze *et al.*, 2002). As previously noted, improvements in log growth in plantations and log use productivity at multiple stages of the supply chain can have a significant impact by increasing plantation log supply and reducing wood usage without decreasing the supply of wood product outputs. This is reflected in Australia's reducing per capita consumption of wood despite growing affluence. To the extent that they reduce demand for logs, improvements in productivity through induced innovation can limit or prevent leakage. The analysis here also shows that while plantation wood sources are often not perfect substitutes for natural forest timbers, changes in technology and processor and consumer preferences can facilitate major and rapid cross substitution effects.

There is limited evidence (in particular due to the presence of other significant drivers) of short term leakage of pulpwood extraction from other Australian states to Tasmania, but stronger evidence of shifts in sawlog production to private natural forest in New South Wales in the middle of the period (that had ceased by the end of the period), and of a small (1.8 per cent) leakage to Western Australian private natural forests later in the period. The latter would make an interesting case for closer assessment in order to understand the role of the relatively unique properties of the Western Australian eucalypt species such as jarrah. The growth in furniture imports is suggestive of weak leakage patterns identified in carbon research (Peters and Hertwich, 2008). Nevertheless, in Australia's case the net whole log equivalent trade balance more than compensated for the increase in log equivalent inflows in furniture (assuming the wood used in the furniture came from international sources). The case for strong leakage in furniture resulting from policy restrictions on public natural forest is, however, contradicted by a number of factors, as described above. Finally, none of the above possible leakage outcomes is supported by evidence that sawlog production from public natural forest has been lower than allowable sustainable yields.

The major volume change in net pulplog exports is supported by a number of previous studies which concluded that a growing international supply of plantation woodchips would lead to a decrease in demand for woodchips from Australia's natural forest (Ajani, 2011; Nelson and Shield, 2003; Pollard, 2006). The historical data presented here confirms these previous analyses.

The results of this analysis are consistent with findings that some countries can 'exhibit net land sparing gains' (Meyfroidt *et al.*, 2010). It supports previous analyses that have suggested wood production from highly productive plantations may compensate for reduced areas of wood production in natural forest (Chomitz, 1999; Murray *et al.*, 2004; Wunder, 2008). At the start of the RFA period in 1996–1997, 49 per cent of Australia's wood was produced from natural forests – by 2016-2017 this had fallen to 14 per cent. Australia's wood production has become increasingly focused on a concentrated plantation estate of about two million hectares (Australia's natural forest estate is estimated at 125 million hectares [ABARES 2018a]).

In Australia's case, land sparing through intensification of wood production from plantations has allowed a significant increase in natural forest conservation with no significant leakage of wood production to other natural forest, either elsewhere in Australia or overseas. Understanding the full potential for land sparing benefits to be realised will also require further analysis of the impacts of creating the plantation estate.[12]

The possible short-term leakage noted above is consistent with the findings of Meyfroidt and Lambin (2009) in that it can take time for intensification effects to replace immediate leakage effects that stem from forest protection policies. It is likely that this delay results not just from the need to establish alternative intensive wood production, but also the time taken for processors and consumers to adjust technologies and demands to utilise the new, often imperfect, substitutes. A critical factor in the Australian experience is that the establishment of the pine plantation estate that effectively substituted for reduced sawlog supply from forest protection policies was put in place some decades before the forest conservation policies of the Regional Forest Agreements.

Substitution towards increasingly low-cost and consistent quality plantation wood raises a question of whether and to what extent market forces and plantation establishment policy alone would have delivered the main conservation benefit of new reserves – reduced logging pressure. Faced with this scenario, forest managers choosing the most efficient harvesting method may have turned to the most productive areas of forest with high conservation values. Harvesting highly productive areas of natural forest may then have enabled the economic viability of natural forest wood production to compete with plantation production in the short to medium-term. In the long-term, costs would inevitably rise as harvesting moved into less productive areas of natural forest. The Regional Forest Agreements are, therefore, likely to have led to a net conservation gain by making areas of natural forest unavailable for wood production and accelerating the pace at which the cost advantage of plantations could be realised. This is similar to the policy logic of putting a price on carbon to avoid the future consequences of climate change.

In popular and political discourse, conservation and industry policy are often assumed to be in tension, if not irreconcilable. The Australian experience described here suggests that a significant degree of alignment can arise between conservation policy and industry policies that support plantation development. Policies that have increased the productivity of plantations have enhanced conservation outcomes, in part by making the production of wood from natural forest less competitive. By restricting access to natural forests, conservation policy has reduced the productive output of natural forest, accelerating substitution towards plantation-grown wood. The complex drivers of policy make it difficult to assess the extent to which this outcome was planned in the Australian case. Nevertheless, the outcome is sufficiently clear to allow a conclusion that plantation development can be a policy enabler of both increased forest conservation and sustained wood production. The key to doing this is to match conservation policy with plantation development via industry policy to ensure an alternative supply of wood with low conservation impacts.

Over recent decades Australia has implemented policies that have transferred public natural forest from wood production to conservation. The resulting reductions in wood production did not lead to leakage of wood production to natural forests elsewhere, and thereby did not undermine the conservation outcomes intended. The key reason for this was the substitution of lost wood supply by increasing plantation wood sources. The Australian experience documented here shows that it is possible for governments to take unilateral action and have a net positive effect on both local and international forest conservation. The experience of Australian forest policy also confirms the potential of land-sparing, in which large areas of natural forest can be conserved by intensifying wood production into plantations with high productivity. Finally, the Australian experience described here provides an active example of the transition of wood production from stewardship forestry over extensive tracts of natural forest to diverging intensive wood cultivation and forest ecosystem management over the natural forest.

Acknowledgements

From *Introduction* this chapter is a revised an updated version of an original paper: Reprinted from Land Use Policy, (52) Russell D. Warman and Rohan A. Nelson, *Forest conservation, wood production intensification and leakage: an Australian case*, pp. 353–362, Copyright © (2016), with permission from Elsevier.

Notes

1 This chapter uses the term *forest conservation* to capture the maintenance of forest ecosystems in a condition that supports their capacity to maintain and enhance the delivery of non-wood values. These include biodiversity and other ecosystem services such as water supply, climate mitigation, carbon storage and cultural values, which can potentially benefit from an absence or reduction in wood production activity.

2 The Australian literature refers to 'native forests', which has the same meaning as the more widely used international term 'natural forest'. The latter is used here. While there is a spectrum of forest conditions between natural and plantation recognised, Australian forests fall nearer to the two ends of the spectrum with clear distinctions between natural forest and plantations. In Australia, natural forests are those of indigenous species that have been established through natural regeneration or regeneration methods intended to mimic natural regeneration processes.

Plantations in Australia are generally intensively managed monocultures of purpose-selected tree species (either indigenous or exotic).

3 In some cases data series do not cover the later few years due to the relevant series being discontinued in recent years or still unpublished.

4 Pulp logs are treated as a residual or arising from sawlog harvest in Australia and generally have not had sustainable yields calculated (National Forest Inventory, 2013). The estimate here is based on the average ratio of pulp logs to sawlogs from public natural forest for 1996–1997 to 1998–1999, which was 55/45.

5 The relative supply impact of a reduction of 2.84 million m^3 of logs in sustainable yield in 2010–2011, represented 0.16 per cent of total global industrial roundwood production of 1,768.9 million m^3 in 2011 and is not considered likely to have affected global prices.

6 Annual per capita wood consumption in Australia dropped from 1.03 m^3 in 1996–1997 to 0.85 m^3 in 2016–2017 (an average annual decrease of 0.09 m^3 a year consistent with the long term trend going back to 1966–1967 when it was 1.27 m^3).

7 Nationally, the area of private land natural forest considered suitable for wood production is much larger than that of public natural forest available for wood production (approximately three times), however the forests on private land are generally of poorer yielding forest types and factors such as owner intentions and access to markets limit the ability of private natural forest to supply logs – see Criterion 1 and 2 for further details in National Forest Inventory (2013).

8 See Figure 2.20 on page 142 National Forest Inventory (2013).

9 Average harvest of 12,700 m^3 in 1996–1997 and 35,500 m^3 in 2006–2007 to 2010–2011. This increase of 22,800 m^3 represents potential leakage of 1.8 per cent of the national natural forest sawlog production over the period.

10 In addition Ximenes and Gardner (2005) reported there was 46,700 m^3 a year of natural forest hardwood timbers used in Australian furniture production in 1999 and noted a large switch from hardwoods and in particular jarrah to plantation softwood in the preceding three years.

11 For example, the portion of New South Wales public natural forest hardwood sawlog harvest being used for 'Joinery/furniture' grew from 1 per cent in 1995–1996 to 7 per cent in 2011–2012 (Wackernagel *et al.*, 2004). Given a halving of public natural forest hardwood sawlog harvest in New South Wales over this time (National Forest Inventory, 2013), this equates to more than a threefold increase in sawlog production for these purposes. The overall decrease in natural forest hardwood sawlog production occurred through a significant decline in its use for structural purposes.

12 It is considered that roughly half of the two million hectares of Australia's plantation estate has been established through conversion of natural forest rather than on previously cleared land, with the bulk of this occurring prior to the introduction of restrictions on conversion in the 1990s (for background see Gerrand *et al.*, 2003; Stephens *et al.*, 1998).

References

ABARES (2018a). Australian forest and wood products statistics summaries: September and December quarters 2017. Canberra: Australian Bureau of Agricultural and Resource Economics and Sciences.

ABARES (2018b). Australian forest and wood products statistics: Exports quarterly index: September and December quarters 2017. Canberra: Australian Bureau of Agricultural and Resource Economics and Sciences.

ABARES (2018c). Australian forest and wood products statistics: Imports quarterly index: September and December quarters 2017. Canberra: Australian Bureau of Agricultural and Resource Economics and Sciences.

Andres, L., Zentner, A. and Zentner, J. (2014). Measuring the effect of internet adoption on paper consumption. *Policy Research Working Paper*. Washington, DC: World Bank.

Atmadja, S. and Verchot, L. (2012). A review of the state of research, policies and strategies in addressing leakage from reducing emissions from deforestation and forest degradation (REDD+). *Mitigation and Adaptation Strategies for Global Change*, 17, 311–336.

Aukland, L., Costa, P.M. and Brown, S. (2003). A conceptual framework and its application for addressing leakage: the case of avoided deforestation. *Climate Policy*, 3, 123–136.

Australian Bureau of Statistics (2018a). 5206.0 Australian National Accounts: national income, expenditure and product, Mar 2018. In: ABS (ed.) *Cat. no. 5206.0*. Canberra.

Australian Bureau of Statistics (2018b). 8752.0 Building activity, Australia, Dec 2017. In: ABS (ed.). *Cat. no. 8752.0*. Canberra.

Burke, B. and Townsend, P. (2011). Agricultural commodities. In: ABARE (ed.) *Forestry [online]*.

Chomitz, K. (1999). Evaluating carbon offsets from forestry and energy projects: how do they compare? In: World Bank (ed.) *Policy Research Working Paper*. Washington, DC: World Bank.

Chomitz, K.M., Buys, P., De Luca, G., Thomas, T.S. and Wertz-Kanounnikoff, S. (2007). At loggerheads? Agricultural expansion, poverty reduction, and environment in the tropical forests. Washington, DC, World Bank.

Clancy, T. and Howell, C. (2013). Sustainable forest management: the Australian context. Canberra: ABARES.

Contreras-Hermosilla, A., Doornbosch, R. and Lodge, M. (2007). The economics of illegal logging and associated trade. Paris: Organisation for Economic Cooperation and Development (OECD).

Dargavel, J. (1998). Politics, policy and process in the forests. *Australasian Journal of Environmental Management*, 5, 25–30.

FAO (2015). Global forest resources assessment 2015: How are the world's forests changing? Rome, Food and Agriculture Organization of the United Nations.

Ferguson, I. (2014). Australian plantations: mixed signals ahead. *International Forestry Review*, 16, 160–171.

Forest Practices Authority (2014). Forest Practices Authority Annual Report 2013–2014 Hobart, Australia: Forest Practices Authority.

Gan, J. and McCarl, B.A. (2007). Measuring transnational leakage of forest conservation. *Ecological Economics*, 64, 423–432.

Gerrand, A., Keenan, R.J., Kanowski, P. and Stanton, R. (2003). Australian forest plantations: an overview of industry, environmental and community issues and benefits. *Australian Forestry*, 66, 1–8.

Guinée, J.B. (ed.) (2002). *Handbook on life cycle assessment: operational guide to the ISO standards*. New York: Kluwer Academic Publishers.

Henders, S. and Ostwald, M. (2012). Forest carbon leakage quantification methods and their suitability for assessing leakage in REDD. *Forests*, 3, 33–58.

Henders, S. and Ostwald, M. (2014). Accounting methods for international land-related leakage and distant deforestation drivers. *Ecological Economics*, 99, 21–28.

Institute of Foresters of Australia (2011). Seeking a new path for forestry in New South Wales. Canberra: Institute of Foresters Australia.

Jaakko Pöyry Consulting (1999). Usage and life cycle of wood products. In: Australian Greenhouse Office (ed.) *National Carbon Accounting System Technical Report.* Canberra: Australian Greenhouse Office.

Kirkpatrick, J. (1998). Nature conservation and the regional forest agreement process. *Australasian Journal of Environmental Management*, 5, 31–37.

Lambin, E.F. and Meyfroidt, P. (2011). Global land use change, economic globalization, and the looming land scarcity. *Proceedings of the National Academy of Sciences*, 108, 3465–3472.

Lane, M.B. (1999). Regional forest agreements: resolving resource conflicts or managing resource politics? *Australian Geographical Studies*, 37, 142–153.

Love, G., Yainshet, A. and Grist, P. (1999). *Forest products: long term consumption projections for Australia.* Canberra, Australian Bureau of Agricultural and Resource Economics.

Macintosh, A. (2013). The Australian native forest sector: causes of the decline and prospects for the future. *Technical Brief No. 21.* Canberra: The Australia Institute.

Markets for Change (2011). No Harvey No: how Australia's largest furniture and electronics retailer is fuelling the destruction of our native forests. Canberra: Markets for Change.

Meyfroidt, P. and Lambin, E.F. (2009). Forest transition in Vietnam and displacement of deforestation abroad. *Proceedings of the National Academy of Sciences of the United States of America*, 106, 16139–16144.

Meyfroidt, P., Rudel, T.K. and Lambin, E.F. (2010). Forest transitions, trade, and the global displacement of land use. *Proceedings of the National Academy of Sciences*, 107, 20917–20922.

Murray, B. C., McCarl, B.A. and Lee, H.-C. (2004). Estimating leakage from forest carbon sequestration programs. *Land Economics*, 80, 109–124.

National Forest Inventory (1998). *Australia's state of the forests report 1998.* Canberra, Bureau of Rural Sciences.

National Forest Inventory (2013). (Montreal Process Implementation Group for Australia and National Forest Inventory Steering Committee). *Australia's state of the forests report 2013.* Canberra, ABARES.

Nelson, R. and Shield, E. (2003). Australian woodchip exports: future commercial benefits of plantation hardwood chip production. *Australian Commodities: Forecasts and Issues*, 10, 74.

Peters, G.P. and Hertwich, E.G. (2008). CO_2 embodied in international trade with implications for global climate policy. *Environmental Science & Technology*, 42, 1401–1407.

Pfaff, A. and Walker, R. (2010). Regional interdependence and forest "transitions": Substitute deforestation limits the relevance of local reversals. *Land Use Policy*, 27, 119–129.

Pollard, D. (2006). Globalisation comes to native forestry. *The Melbourne Review: A Journal of Business and Public Policy*, 2, 61.

Rapidel, B., De Clerk, F. and Beer, J. (2011). *Ecosystem services from agriculture and agroforestry: measurement and payment.* Oxon: Earthscan.

Schwarze, R., Niles, J.O. and Olander, J. (2002). Understanding and Managing Leakage in Forest-Based Greenhouse Gas Mitigation projects. *Philosophical Transactions of the Royal Society of London. Series A: Mathematical, Physical and Engineering Sciences*, 360, 1685–1703.

Simon, H. A. (1972). Theories of bounded rationality. In: McGuire, C.B. and Radner, R. (eds.) *Decision and organization*. Amsterdam: North-Holland Publishers.

Sohngen, B., Mendelsohn, R. and Sedjo, R. (1999). Forest management, conservation, and global timber markets. *American Journal of Agricultural Economics*, 81, 1–13.

Stephens, N., Tickle, P. and Sun, D. (1998). *Plantation potential studies in Australia: an assessment of current status*. Canberra, Bureau of Rural Sciences.

United Nations (2018). UN Comtrade Database. In: Nations, U. (ed.). New York.

URS (2007). Australia's forest industry in the year 2020. Canberra: Department of Agriculture, Fisheries and Forestry.

Wackernagel, M., Monfreda, C., Schulz, N.B., Erb, K.-H., Haberl, H. and Krausmann, F. (2004). Calculating national and global ecological footprint time series: resolving conceptual challenges. *Land use policy*, 21, 271–278.

Whittle, L., Hug, B. and Burns, K. (2012). Leakage from avoided harvesting in native forests under the Carbon Farming Initiative: a qualitative assessment. Australian Bureau of Agricultural and Resource Economics and Sciences.

Wunder, S. (2008). How do we deal with leakage. In: Angelsen, A. (ed.) *Moving ahead with REDD: issues, options and implications*. Bogor, Indonesia: Center for International Forestry Research.

Ximenes, F.A. and Gardner, W.D. (2005). Production and use of forest products in Australia. In: INDUSTRIES, FRRNDOP (Ed.) Technical Paper. Sydney: Forests NSW.

6 A social turn

Decentralisation in Indonesia – a forestry history

Our attention now shifts to the social (and the political), where institutions of the social-ecological system can be analysed and understood. The case study in this chapter focuses on the role of institutions of forestry in Indonesia, a country with a complex history at the intersection of many key global processes such as modernity, colonisation, globalisation and industrial development. The emphasis here is on the governance and social and political dimensions of wood and forest social-ecological systems. Power, as a social phenomenon, has been significant in the processes of wood and forest social-ecological system resilience and transformation. Analysing the social dimension by following the threads of power is the second 'tack' of Latour's tripartition of critique for complex networked systems.

Introduction

This chapter gives particular attention to the role of centralisation in governance and forestry in order to study how power has been influential in shaping wood and forest social-ecological systems. Centralisation in governance and forestry had been widely written about in the several years after Indonesia's change to decentralised governance, in the years immediately after 1998 when President Soeharto was forced to stand down after three decades of his New Order regime. However, several years have now passed since that flurry of academic interest and this chapter reviews this process with the advantage of a little more distance in time.

The question of centralising and decentralising governance is of considerable relevance to the unfolding of the greater global transition of wood production where a range of processes have taken place involving shifts of power between centres and peripheries, such as that of the industrial core and its periphery, the dissemination of stewardship forestry from the colonial core, and the rise of the south as a force in global wood production.

As governments and other policy actors grapple with forest governance issues, so, too, the forest use and wood production systems that are governed are evolving. Changing socio-political-economic processes related to the development of industrial societies have created distinctive patterns of forest use in the lead up to, during, and after, rapid industrialisation. This is the process of transition from basins of forest exploitation to stewardship forestry and then to wood cultivation and forest ecosystem management. Indonesian wood and forest social-ecological systems have moved through these basins rapidly in the last half-century, with basins overlapping in various ways. The forest exploitation basin, with land clearing, indiscriminate extraction of high value species, illegal logging resulting in part from poor property right allocations, and little or no active forest management, has been observed extensively in Kalimantan, Sumatra and Papua. At the same time, stewardship forestry institutions have grappled with containing this process, well aware of its unsustainability. And even as this occurs, wood cultivation, particularly in the form of industrial plantations, has begun to rapidly develop.

As well as being expressed through these land use patterns, the basins can also be seen in terms of their institutional expression. Forest exploitation generally does not have formal institutions, being associated with frontier exploitation. Indonesia's expanding populations have taken advantage of new demand for agricultural land and wood from what is often treated as an expansive commons, and is often conducted with little or no central government control, or modern legal or tenure frameworks. On the other hand, where there are indigenous or long standing local populations there is likely to be a range of customary or traditional practices and tenures. Stewardship forestry evolves in response to increased demand for stronger institutions of control, law and law enforcement in order to restrain the exploitation. This has included assertion of central government control over large areas of forested land and, thus, the emergence of forestry as a form of leviathan governance responding to what is seen as an unfolding tragedy of the commons,[1] but also aligned with powerful interests in society taking control of resources. The institutions developed by the central government at this stage are often imposed over pre-existing customary institutions and tenures of ownership. However, within this basin of stewardship forestry management of forests becomes contested. Institutions competing with those of forestry include civil society, protected area management agencies and revitalised traditional owner interests, especially in the extensive remaining areas of natural forests at and beyond frontiers of wood extraction. At the cleared/ developed land interface there is also an increased blurring of forestry and agriculture as wood cultivation becomes established in lands previously cleared of forests. At this point a series of spheres of interest[2] in forest management and ownership have developed that are international, national and local. They have developed in a context that is driven by a struggle for access to and control over political power and resources. Figure 6.1 provides a schematic overview of the contested spheres of interest paradigm. The following history expands on the nature of the contest between these spheres through a review of the shifts between centralising and decentralising governance.

In 1999, in the aftermath of the fall of the presidency of Soeharto, the Asian financial crisis and major droughts and fires, Indonesia passed new laws that decentralised substantial responsibilities, powers and finance from the central

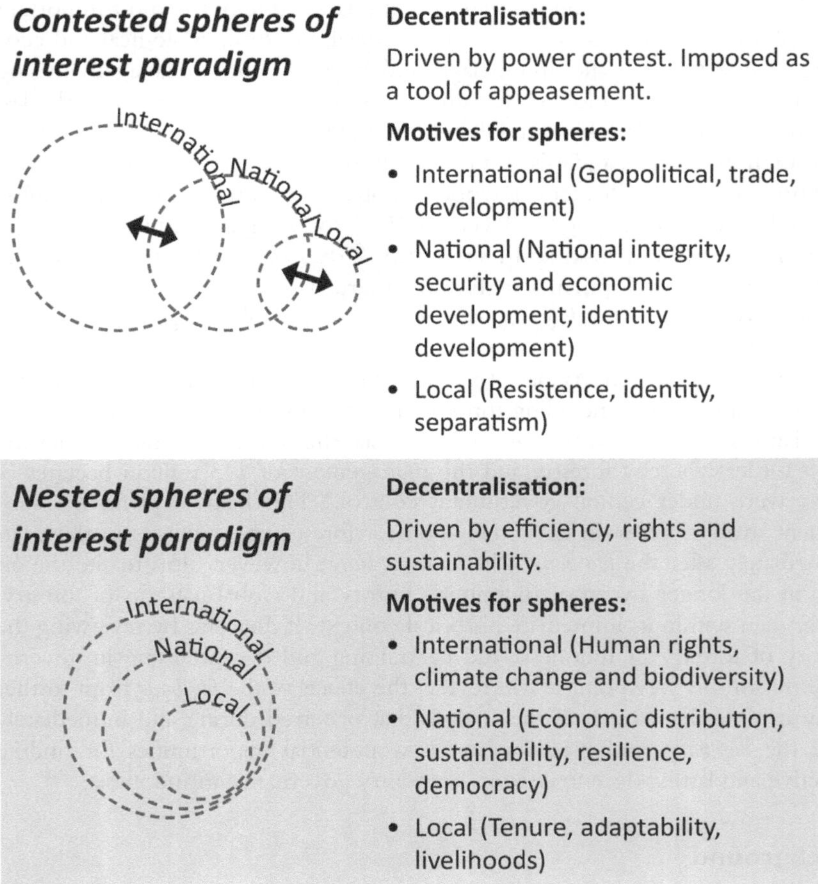

Figure 6.1 A schematic of the spheres of interest, including drivers, relationships and the motives of individual spheres.

government (Colfer and Resosudarmo, 2002). It has been cited as one of the most significant decentralisation efforts undertaken in recent times (Crouch, 2010). Indonesia has a long history of tension between central interests, generally based in Java, and the other islands of the archipelago. With forestry being a significant part of Indonesia's economy, the evolution of forestry institutions has featured in these core/periphery relationships. There is a considerable body of literature on the impacts of the decentralisation of government in Indonesia in the *reformasi*[3] period. In particular, the impacts of decentralisation in forestry (Barr *et al.*, 2006a; Colfer and Resosudarmo, 2002; Larson, 2005; Moeliono *et al.*, 2009; Palmer and Engel, 2007), and subsequent recentralisation (Bullinger and Haug, 2012; Crouch, 2010), have been richly reported.

The use of forests and the growth of forestry have been central to the development of the Indonesian state. Indonesia has one of the most important areas of forest in the world; of significant social, political, ecological and economic value to both the archipelago and beyond. Forest management and wood production in Indonesia has undergone significant changes over the last few decades (Obidzinski and Chaudhury, 2009). Deforestation rates have been high by world standards and peaked in the years 1997 to 2000 (Ministry of Forestry, 2009). Plantation wood supply, as a portion of total pulplog demand, rose from negligible levels in 1999 to around 20 million m^3 a year ten years later (Obidzinski and Dermawan, 2012; see also Figure 6.2). In the midst of these developments in forestry there has been dramatic change of forestry governance; the *big bang* of decentralisation following the end of the New Order.

As noted earlier, forestry has played a key role in colonisation and the emergence of nation states including through assertion of power by central interests over land and people. In the case of Indonesia, this led to conditions under the New Order whereby forestry and the management of 143 million hectares of forest were under central government control.[4] This helps explains the subsequent strong interest in decentralisation for forest policy analysts in the years immediately after the *big bang*. The analysis here, however, aims to site the *big bang* in the longer sweep of Indonesia's history and global patterns of forestry, rather than within its immediate historical context. It does this by reviewing the history of forestry in Indonesia and centralising and decentralising in governance. From this viewpoint it will review the effects of the *big bang* from further away in time than much of the analysis that occurred during and immediately after the *big bang*, and speculates on new potential opportunities for finding effective and lasting decentralisation in forestry governance in Indonesia.

Background

Decentralised governance is distinct from federal systems (for example, Australia, Malaysia and the US) where delineation of powers between a federal government and its constituent parts is clearly spelt out in a constitution (Gregersen *et al.*, 2005). By contrast, Indonesia is a unitary state where formal powers reside with the centre. Decentralisation refers to the shift of powers and responsibilities from a centralised government to components of its jurisdiction. These powers/responsibilities can be *political* (making decisions), *administrative* (managing allocated resources and responsibilities), *fiscal* (taxation and revenue raising), and *market* (privatisation and deregulation of private industry) (Blaser *et al.*, 2005). In Indonesia these responsibilities can shift from the central government to provincial, district or village levels, as well as through shifts in ownership and rights to customary custodians and private firms (Ferguson and Chardasekharan, 2005).

Decentralisation of governance in developing countries was advocated by developed countries and international development agencies in the 1950s and

1960s. This was strongly associated with colonial powers seeking to establish what they considered were the good governance benefits of sharing power with local government jurisdictions; specifically, as modelled on governance structures that existed in developed countries (Conyers, 1983). In the late 1970s and 1980s it emerged again as a 'fashionable' topic, especially in international development organisations (Conyers, 1983). This time there was a stronger academic basis and it was seen in particular as a vehicle to further rural development in developing countries (Ahmad and Talib, 2011).

Despite this longstanding history of advocacy there is still considerable uncertainty about the advantages and disadvantages of decentralisation (Moeliono *et al.*, 2009). Examples of adverse findings include increased deforestation stemming from increasing subdivision of districts (Burgess *et al.*, 2012), decreasing economic equality as a result of poorer regions lacking capacity or resources to take advantage of the opportunities potentially offered by decentralisation (Rodríguez-Pose and Ezcurra, 2010), disappointing gaps between goals and actual outcomes, especially when a top down approach has been applied (Wunsch, 2013) and mixed outcomes for local communities subjected to changes in centre/local control of resource rents (Bullinger and Haug, 2012). An overarching review of decentralisation research in relation to forestry in Indonesia was also mixed in its findings (Barr *et al.*, 2006a).

Forests are often used as common-pool resources, with a number of influential models explaining the problems they suffer when exploited as such. Elinor Ostrom (1990. pp. 2–7) notes three models that describe core problems with common-pool resources. There is the 'tragedy of the commons' (Hardin, 1968), in which lack of ownership leads rational players to maximise their use of the resource, leading to overexploitation. The decision making of actors in this tragedy is clearly evinced in the second model, the prisoner's dilemma. In this model each agent acts rationally to best their own outcomes, but will contribute to an overall worse outcome for everyone. Finally, Ostrom refers to the work of Olson (1965) in which Olson finds that actors will shirk their fair contribution to a collective action if they will benefit anyway – free riding – such that if all, or many, actors choose this option, the collective action fails, again leaving everyone worse off. However, Ostrom finds that these models have limitations, especially in how they come to be applied to policy solutions. She notes the existence of a body of work pointing to only two outcomes. One, the 'leviathan' position, takes the view that an overarching central government can see the problem from above the self-interest of individuals. Such a government will use rational decision making with all available information to assert an optimal outcome. The second position advocates privatisation. Here, clear private ownership removes the problems of the common-pool resource. This leads to optimal efficiency of resource use through the discipline of the market. Though often framed as opposite approaches, both positions require an external agent of control (Ostrom, 1990), typically a central government. Despite this, however, there is evidence that self-organised community governance is an effective mode of common-pool resource management (Agrawal, 2003; Ostrom, 1990; Wunsch, 2013). Given that such self-governing systems are possible, then, this becomes a potential third way – one where the actions of the external central agent can support the conditions necessary for decentralised governance. This would include giving jurisdictional rights to a local or decentralised level of governance, but could well require other actions such as building the necessary supportive institutions, both at central and local levels.

History of Indonesian (de)centralising relations in forestry

The colonising Dutch struggled over a number of centuries to assert control over the archipelago (Vlekke, 1959). Only in the early twentieth century did they reach a point where they were able to tenuously control most of what is now Indonesia and imagine it as a territorial unit (Ricklefs, 2001). During this time there was a concurrent expansion in forestry control, which was entwined with the territorial mission of the colonists (Galudra and Sirait, 2009; Peluso, 1992; Peluso and Vandergeest, 2001).

Early Dutch forestry interests were focused on the naturalised teak resources of Java (Pandey and Brown, 2000). It was here that the strongest instruments of central state control were progressively instigated. This started in the early nineteenth century and culminated in the Forestry Law of 1927 that gave the colonial Forest Service control over all forests on Java (Peluso, 1992).[5] However, the Dutch still struggled to establish control over forests on the Outer Islands (Galudra and Sirait, 2009).[6] At this stage the forests of the Outer Islands were of little forestry-related economic interest, and largely controlled by traditional ownership patterns. Rulings in the colonial judiciary and a spirited debate about the legal rights of the state versus traditional ownership of forests beyond Java (Peluso and Vandergeest, 2001) possibly kept expansionary impulses at bay. This state of affairs set the pattern for forestry in Indonesia at independence in the aftermath of the Second World War – a highly centralised Javan forestry and a high degree of traditional control in the Outer Islands.

The new Indonesian Forest Service maintained key parts of the personnel, culture and structure of the Dutch colonial Forest Service (Peluso, 1992) and there were no changes to the forestry laws or regulations of the Dutch during the period of Japanese occupation or subsequent first decade of independence (Kartasubrata, 1985). The inherited residual power of provincial governments on matters of forestry was clarified in 1957 when the then President Soekarno moved to ease concerns in some provinces that the Jakarta government intended to claim control of the natural resources of the Outer Islands (Barr *et al.*, 2006b). This regulation spelled out that control over issuing concessions resided with the provinces (outside Java and some limited parts of Kalimantan, under the control of the national Forest Service). It was an act of decentralisation, particularly of administrative responsibilities. Despite this, it also required that permits be issued and management decisions of provincial forest managers to be consistent with the centre's regulations (Kartasubrata, 1985).

As before, *adat* – customary or traditional practice and law – legal systems continued to be recognised throughout the Outer Islands. In contrast, on Java the lack of recognition of these, based on the central Forest Service's complete ownership of Java's forests, was a source of considerable tension (Barr *et al.*, 2006b). In 1960 the Basic Agrarian Law aimed to clarify the dual national/*adat* legal system in relation to land ownership and control. It was influenced by the Indonesian Communist Party and intended to extend and preserve rights of the population to the use of the land (Peluso *et al.*, 2008). However, like the 1957 forestry regulations, it contained a measure that simultaneously entrenched central government control – a caveat that land was controlled by the central government where it was in the 'national interest'. In any conflict between regional interests or central (national) interest, a critical veto on power sharing remained with the centre.

At this time there was also growing international pressure for the Indonesian Government to promote economic growth through the exploitation of its natural resources (Barr *et al.*, 2006b), including forestry.[7] The Soekarno government took its first steps in this direction establishing some logging concessions on Kalimantan in the 1960s (Barr *et al.*, 2006b). At this stage there was still a reasonable degree of consideration for *adat* in forests in the Outer Islands. It is likely that this reflected a lack of technological means, and commercial potential, to develop the hardwood resources of the natural forests of the Outer Islands. That changed dramatically with the start of the New Order.

A year after Soeharto commenced the New Order, the Forestry Law of 1967 was passed. It established central control over 143 million hectares of forest across Indonesia (Barr *et al.*, 2006b). Subsequent regulations gave the central government the ability to issue large logging concessions over this area, while provinces continued to be able to also issue licences. For this period, called *Banjir Kap*,[8] there was a rapid expansion of logging, characterised by much confusion and poorly controlled exploitation. In 1970, Soeharto issued Regulation 20/1970, outlawing *Banjir Kap* and giving the central government sole rights to issue logging concessions, which had to be bigger than 50,000 hectares (Ross, 2001). The central government did this rapidly. Indonesia went from a country that exported 209,000 m^3 of wood in 1965 to being the world's largest tropical wood exporter in the late 1970s, exporting more than 20 million m^3 of unprocessed logs a year (Barr *et al.*, 2006b). In this period the two main policy solutions to common-pool resource problems were fully implemented – leviathan control over forests, in partnership with a privatised system of concessions over the forest estate.

In the early 1980s a total ban on the export of whole logs was instituted to encourage domestic processing (Hyde, 2012). This led to consolidation of the nation's wood industry into a much smaller number of larger companies dealing directly with the central government. This served to further consolidate central control over forestry (Barr *et al.*, 2006b). The boom in whole log exports ended but was quickly replaced by a boom in plywood export. Then in the 1990s a pulp and paper boom followed (Figure 6.2). Each of these waves of wood processing – whole logs, plywood, pulp and paper – placed Indonesia as one of the world's largest exporters in each category at the time. Wood and wood products were a major source of export earnings for the country. Each of these subsequent waves also involved a shift to more capital intensive forms of wood processing. More capital intensive wood processing involved greater degrees of foreign investment and ownership, larger corporate players, increased wealth distribution issues within Indonesia, and increased geopolitical and trade considerations – all areas of responsibility to justify (require) Jakarta's central control. Further, in the political contest for resource and wealth control, the bigger players and sums of money involved would also have served as incentive for increased central government control.

In the late 1980s there was a realisation that the rate of natural forest exploitation was unsustainable. A plantation establishment regime to provide a long-term sustainable wood supply was started, wherein wood companies were given access to areas of mostly degraded forest for conversion to plantation (Barr *et al.*, 2006b). This was in response to unbalanced wood industry development policy over previous decades that involved subsidised and uncontrolled expansion of processing capacity, requiring an unsustainable rate of harvest from Indonesia's natural forests. Consequently, the wood products sector went into deep decline in the 2000s, with nearly one-third of wood products sector jobs going in the period 2001 to 2006 alone, and 2004 to 2006 seeing a 55 per cent decline in plywood export and a 40 per cent decline in sawn timber exports (Obidzinski and Chaudhury, 2009, p. 80). In addition, permits to convert forest to palm oil were also issued on a large scale in the 1990s (Barr *et al.*, 2006b). Combined with illegal logging, these

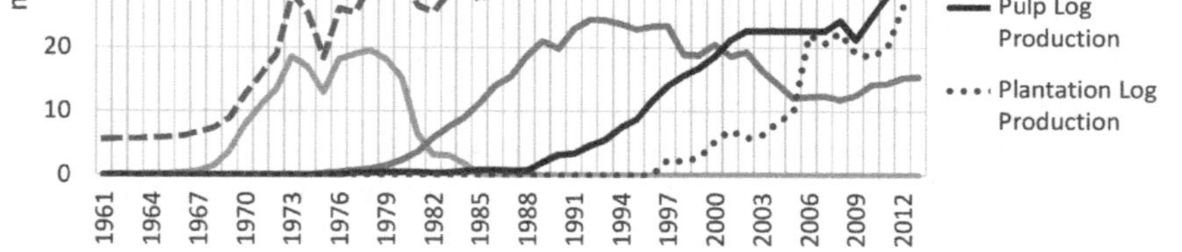

Figure 6.2 Production of total logs and the three successive waves of log uses.[9]

processes contributed to Indonesia having one of the world's highest rates of deforestation through the 1990s. There was a growing recognition of serious deforestation, industrial overcapacity in the wood processing sector and unsustainable logging rates (Barr *et al.*, 2006b).

This forestry crisis was unfolding as the New Order started to collapse, both in the wake of the Asian financial crisis of 1997 and a downturn in support for Soeharto. There was considerable concern that Indonesia was on the verge of becoming a failed state. In response to this crisis, the government of President Habibie, who replaced Soeharto in 1998, moved quickly to decentralise power (Nordholt and Van Klinken, 2007). There was a calculation that this was needed to stave of disintegration of the nation in the face of growing separatist agitations, part of which involved demands for greater access to natural resource wealth within the regions (Barr *et al.*, 2006c). Simultaneously, decentralisation allowed a shift of power to the regions, where the hegemony of Soeharto's long- ruling Golkar party was sufficiently well-established to be thought resilient to the potential loss of power facing the national government through proposed fair elections (Crouch, 2010). At the same time as these internal pressures were occurring within Indonesia, international development policy and, specifically, trends in international forest policy were placing greater emphasis on the value of decentralisation as a means to increase 'efficiency, equity and democracy' (Larson, 2005, p. 3). International development agencies and forestry organisations promoted decentralisation of financial and governance structures for communities living in forested regions (Blaser *et al.*, 2005). In its negotiations for structural reform after the Asian financial crisis, the International Monetary Fund also added indirect pressure for decentralisation in Indonesia (Barr *et al.*, 2006c). These pressures combined to see the implementation of 'one of the developing world's most decentralised systems' (Crouch, 2010, pp. 8–9).

The period of decentralisation was one of intense competition for influence. New coalitions were formed between local interests seeking regional advantage from the diffusion of power away from the centre (Kimura, 2010). Forestry governance was left in a confused state as a result of unclear divisions of responsibility between the centre and the regions. Indeed, it was observed that '[n]owhere has the disorderly and sudden nature of the changes brought about by decentralisation been more apparent than in the management of Indonesia's forestry resources' (Resosudarmo, 2003, p. 230). Despite the passing of *Otda*[10] laws (Law No 22) that devolved considerable forestry management from the centre, it was only a few months later that the new Forestry Law[11] of 1999 was adopted. This acted to re-establish central control over forests and their use (Siswanto and Wardojo, 2005). However, it created an environment in which conflicting laws created legal ambiguity (Arnold, 2008) that led to numerous demarcation disputes between various levels of government and private businesses (Crouch, 2010).

Regulations passed soon after the 1999 Forestry Law assigned district governments the rights to issue small (<100 ha) wood permits. John McCarthy describes the period of confusion between 1999 and 2003 in the province of Central Kalimantan (McCarthy, 2007) where an increased level of lawlessness in logging reigned – there was a proliferation of armed pirates and protection gangs on the main rivers – until a procession of central government regulations reinstated central control. This chaos is reflective of the period of *Banjir Kap* that followed the Forestry Law of 1967, described earlier. In both cases the response to confused enactment of newly implemented dual decentralised governance systems in an environment of radical change was to move power back to the centre. Two examples of this shift from the *Otda* period were a series of central government moves to rein in the proliferation of small scale wood licences being issued by the regions, culminating in the regulations of 2002 which shifted responsibility back to the centre, and a decree in February 2001 affirming that ministerial orders (central government) overruled any regional regulations where they conflicted. By 2002 forestry was being rapidly recentralised (Barr *et al.*, 2006d).

Central control was more generally reconsolidated with the passage of Law 32/2004. This clearly spelt out that the central government delegated powers to the regions, but also included measures such as the centre's veto right over regional regulations and budgets, effectively returning key powers to the centre. The laws of 2004 marked the end of the 'high season of regional autonomy' (McCarthy, 2007). However, in exchange for this increase of central oversight, and in clear recognition of the idea of a unitary state, the laws of 2004 also set up the process for direct elections of provincial governors and district heads. This built in a counterbalancing long-term commitment to representation, accountability and acknowledgement of the legitimacy of regional governance (Crouch, 2010). The overall effect of the 2004 laws was to restore to the centre power and responsibility in matters administrative and fiscal, but in exchange for greater political decentralisation.

Despite considerable attention focused on the revival of *adat* in Indonesia in the period of *reformasi* (Davidson and Henley, 2007), its application as a part of central government forestry decentralisation since the *reformasi* has been weak. This is consistent with a shift in the rationale for decentralisation from maintaining culture and ethnic identity to economic efficiency (Rodríguez-Pose and Sandall, 2008). Through the *reformasi* period there were examples of logging companies being required to work with local communities as well as increased local community actions in support of *adat* and customary tenure claims (Barr *et al.*, 2006e). In 2006 regulations provided for allocations of Indonesian forests to community-based forest management, however some years later such measures were still considered to remain at the margins of forestry policy despite clear evidence of their job creation potential (Royo and Wells, 2012). Perhaps most significantly for the revival of *adat* and customary tenure, in 2013 the Indonesian Constitutional Court found that the 1999 Forestry Law did not override pre-existing customary tenure and *adat* in state forests (Johnson, 2013). The ramifications of this are far-reaching, but slow to be realised as they are complicated by resistance and ongoing legal, mapping and institutional ambiguity (Myers *et al.*, 2017). The court's finding is likely to impact ownership of millions of hectares of forest – in 2015 the government moved to reallocate 12.7 million hectares of forest to local forest communities while leading indigenous rights groups talk about making claims for tens of millions of hectares (Astuti and McGregor, 2017).

The above shows that over the course of Indonesia's development, assembling and controlling the territory of Indonesia was a strong centralising force – the long history of Indonesia is one of increasing centralisation consistent with the concentration of power into the nation state. Forest use and management have played a significant part in this. However, as shown in the discussion above there has been some sensitivity to the limitations of centralisation and subsequent efforts to decentralise have also occurred (often in a cyclical fashion).

The effects of decentralisation on governance in forestry

The review above suggests that application of decentralisation to date, including the *big bang*, was primarily an instrument to further territorial control during periods of weakened central government by appeasing separatist and regional agitations. Demand for decentralisation appears in the colonial discourse in the early twentieth century (Vlekke, 1959) in response to the emergence of the centralised state. Dutch attempts to create a federation at the end of their rule was seen as a ploy to keep a level of power and influence in the regions that would not be possible in a unitary state of Indonesia alone (Crouch, 2010). Decentralisation, then, has often been used as a tool of power wielded by a central authority to maintain control. This use of decentralisation helps explain the initial failures of the *big bang* in forestry. At a governance level, for example, assignation of responsibilities to districts rather than to the provinces, which were more strongly associated with separatist and identity based struggles (Ribot *et al.*, 2006), is an obvious source of potential failure. At the district level, a lack of capacity to implement good forest governance contributed to problems that gave justification for the centre to reinstate its control.

Overall, there was poor implementation of *Otda* law (Arnold, 2008), and 'weak governance structures' in general (Gregersen *et al.*, 2005). The idealistic drivers of reform struggled to be realised as they collided with Indonesian power conflicts and the essentially political motives behind *Otda*. Vedi Hadiz notes that 'entrenched local predatory interests' effectively co-opted externally driven efforts to impose good governance for their own ends (Hadiz, 2010). The extension of 'money politics' to the periphery was reflective of this failure to realise the aspiration of decentralisation as a tool to undo the corruption of the previous centralised regime (Crouch, 2010), and despite the democratic promise of decentralisation it acted to decentralise corruption as well (Malley, 2003). At the end of the New Order the Indonesian state 'had become the possession of its own officials' (Hadiz and Robison, 2013). It follows, then, that the *Otda* which transferred those officials to the regions simply rearranged the geography of the ruling class. It failed to undo their control over the state. Fukuoka (2013) describes how, when a 'sultanistic regime'[12] democratises after its collapse, it can still be prone to the same clientelistic patterns of power relations. In this sense, like corruption being decentralised, the patron-client relationship of the sultanistic regime was decentralised. Likewise, the oligarchic nature of Indonesia under Soeharto was continued in spite of *Otda* (Hadiz and Robison, 2013). In this, then, the promise of the radical state institutional rearrangement of *Otda* struggled to be realised, because it was prone to the characteristics of the preceding centralised system.

The radical changes of *reformasi* also failed to address issues of poor property rights. These had been in limbo since the early promise of fairer land distribution in the Basic Agrarian Law had been brushed aside in the New Order period, tarnished by the law's initial association with the Communist Party (McCarthy, 2000). Control of forested land by the centre under the Forestry Law of 1967 allowed the Soeharto government to transfer rights over forests to a small number of large private interests closely aligned with Soeharto himself, while using the discourse of leviathan forestry as justification (McCarthy, 2000). The long history of increasing central control of Indonesia's forests by the institutions of professional forestry and centralised government corresponded with a decrease in local access to, and use of, *adat* (Peluso and Vandergeest, 2001). In addition, traditional systems of forest management were also vulnerable to breakdown in the face of rapid social and technological change (Dietz *et al.*, 2003). The disruptive influence of colonisation and central government appropriation of traditional forests and their use from customary owners can lead to the 'tragedy of invasion' (Brightman, 1987). In Indonesia's case, the tragedy continued after the colonial invasion ended.

In the contest between *adat* property rights and those of the state, demarcation of forests through mapping has also been essential to the assertion of territorial control by central government forest agencies (Peluso, 1995). The Dutch first used mapping as a tool for control of Java's forests in the nineteenth century and subsequently the Forestry Laws of both 1967 and 1999 asserted control over extensive forests through their associated maps. In May 2014, in a speech in Jakarta, President Susilo Bambang Yudhoyono noted the current unified Indonesian mapping project (Yudhoyono, 2014), designed to address widespread inaccuracy and contradiction in existing maps in order to provide greater clarity to central government forestry initiatives along with increased transparency.[13] The refinement of mapping and clarification of land tenure will improve the ability to undertake community based forestry initiatives (Royo and Wells, 2012), allowing this new mapping to contribute to 'counter-mapping' – using maps to build counter-claims to the past alienation of territory by central authorities (Peluso, 1995) although the imbalance of mapping power continues to make progress challenging (Radjawali and Pye, 2015). The emergence of counter-mapping initiatives also indicates increased central government responsiveness to local and regional needs, including *adat*. This, along with recent changes in the legal standing of customary rights under the Forestry Law and other initiatives in community based forestry, point to a significant reshaping of the relationship between the spheres of interest. Here, it is speculated that a new paradigm for the relationship between spheres of influence is emerging – one that is based on efficiency, rights and sustainability rather than the power contest of the old paradigm (Figure 6.1).

The imposition of central government control over forests is consistent with a strong historical global current in the institutions of forestry. These were ostensibly developed to protect forests from overexploitation in a tragedy of the commons scenario, but often emerged as a result of power contests over use of strategic resources, including wood, hunting and land (Sayer *et al.*, 2005). This alignment between forestry institutions and leviathan government has been mirrored in leviathan tendencies in the modern conservation movement towards protecting natural areas (Sayer *et al.*, 2005). This is consistent with ongoing impositions of centralised control over forests through mechanisms such as REDD and its related moratoriums on the issuing of new logging licences (Murdiyarso *et al.*, 2011). However, there are also calls for decentralisation as an optimal method of delivering REDD+ (Karsenty and Ongolo, 2012), including the need to recognise local and traditional rights (Wright, 2011). Such an approach would act to align local spheres of interest with national and international spheres. This reflects the possibility that new motives operating in the international sphere of interest could push a more nested set of relationships between the various spheres of interest. Here, a shift could occur, from spheres of interest competing for resources and political power, to spheres of interest cooperating for mutual resilience (Figure 6.1). Drivers for the emergence of these more cooperative arrangements between the spheres of interest are increasing recognition of the importance of extensive and interconnected social, economic and ecological systems, new governance responses to this understanding and related shifts to delivering efficiency, rights and sustainability. Implementation of international development initiatives such as REDD+ and the Sustainable Development Goals (Sachs, 2012) are potential examples of such an evolution.

In terms of evolving relationships between spheres of interest, unfolding changes in forestry could also provide opportunities for increased decentralisation. The growing volume of wood coming from plantations in Indonesia (Figure 1.3) could reduce pressure on natural forests for wood extraction. As noted earlier, this has been suggested as a means to allow greater use of natural forests for such ecosystem services as biodiversity, where previously this conflicted with wood production. By the same logic, it could also potentially reduce conflict between revitalised *adat* forestry and the large commercial drivers behind concession-based natural forest logging by shifting the bulk of the timber demand to more concentrated areas of intensive timber cultivation. This would allow extensive natural forests and adjoining agroforestry landscapes some respite from the centralising demands of highly concentrated and capital-intensive timber industries. This structural change in forestry could increase the prospect of a new paradigm of nested, or cooperative spheres of interest to emerge.

An increasingly resilient sense of Indonesian nationality could also provide the central government with more capacity to relax its control over forests (territory), providing another opportunity for increased local control where that is consistent with best outcomes. Finally, the arrival of a more democratic society in Indonesia has allowed the emergence of new coalitions urging the localisation of power over natural resources. For example, environment movements and local political interests have allied against the central government – despite clear differences in their views – to seek a shift in decision making away from the centre (Kimura, 2010). Also, new alignments between REDD, the One Map Initiative and the constitutional court decision are all changing the dynamic for decentralisation of control over forests, although not without complications (Astuti and McGregor, 2017). This exploration of alignments between different spheres of interest provide opportunities for the discovery of synergies that might facilitate cooperation between spheres of interest rather than contest. It is possible, in this environment of cooperative spheres of interest, that a lasting and effective decentralisation can emerge. That is, in complex interconnected social-ecological systems decentralisation has a supportive logic based on building system resilience for the long term delivery of efficiency, rights and sustainability. By contrast, in a contested spheres of interest paradigm dominated by power relationships it is likely that the interests of the national sphere will remain separated from local sphere interests and tend to draw power to itself, in the logic of maintaining power.

Decentralisation into the future

In relation to Indonesian forestry, the phrase *big bang* is something of a misnomer. Over the longer sweep of Indonesian history the Indonesian state has been mirrored by the rise of increasingly centralised forestry governance, albeit with fluctuations due to periodic decentralisation efforts. Even in the brief period between 1999 and 2002, forestry was much more centralised in Indonesia than at any time prior to independence, with a centralised bureaucracy having far-reaching powers across the archipelago. While parts of Indonesia had relative decentralisation in forestry during some of this history (for example, in the years after the 1957 regulations there was a parallel forest bureaucracy in the Outer Islands) these periods were generally associated with little or no commercial or strategic interest in the use of forests. Where commercial or strategic value developed there has been an associated move by the centre to quickly claim rights and responsibilities over relevant forest resources, notably in the regulatory release and recapture of power that occurred in the early years of the New Order and 30 years later in its aftermath.

In the case of the *big bang* the reality of decentralisation of forestry in Indonesia, in the intensively studied years after 1999, does not support the idealised benefits of greater equity, increased wealth and more sustainable forest management. However, this can plausibly be explained through the poor implementation of *Otda* in forestry; it was far from comprehensive and rapidly and substantially reversed. Regardless, the major forest allocation decisions relating to the predominant large-scale wood concessions and related capital-intensive wood processing facilities, along with control of forests for carbon and biodiversity conservation purposes, has essentially remained with the central government. Despite this, there is evidence of an emerging decentralised forestry potential through increased support for customary forest management. In addition, reductions in corruption and increasing democratisation could raise the ability of local spheres of interest to be more effectively involved in forest related decision making. This is critical to realising the benefits of decentralised and *adat* forestry in Indonesia, where many of the rural poor live with forests.

In 1983 Diana Conyers asked if the emerging wave of interest in decentralisation was a repeat of earlier phases in the 1950s and 1960s and, as such, a reflection of a periodic pendulum swing between the two, or a result of a more fundamental change in governance (Conyers, 1983). In the case of Indonesia and forestry the *big bang* reforms contain elements of both. Shifts between centralisation and decentralisation have continued to occur, especially in response to contests over power between the centre and regions, reflecting the pendulum effect – Siswanto and Wardojo (2005.p. 74) note that '[s]ince independence in 1945, the governance system has changed no fewer than six times, back and forth from centralised to decentralised.' This could be seen as a logical outcome of a contested spheres paradigm. However, given the complete centralisation that occurred under the Soeharto regime, it also seems likely that, providing the democratisation of Indonesia continues with a higher degree of local participation in decision making, the emerging decentralisation measures may reflect a fundamental and lasting change, with potential for greater say and benefits going to local forest communities.

Forest ownership in Indonesia, as in much of the world, continues to be dominated by central government. However, also replicating global patterns, control of government-owned forests is changing in response to demand for increased local and traditional management as well as independent certification (Agrawal *et al.*, 2008). There is also a need to recognize the presence of ongoing structural evolutions in forestry – these require governance that is adaptive to changing conditions (Dietz *et al.*, 2003). By allowing greater flexibility between formal ownership, as determined by Indonesia's forestry laws, and control and governance of forests, there is increased scope to find a more nuanced governance structure suited to the demands of Indonesia's forests and dependent communities, with polycentric spheres of mutually supported interests. This means that the promises of decentralised forest governance might be delivered, not as a central government-ordained *big bang* – but, rather, as a progressive, paradigmatic evolution.

On a final note, coming back to the central idea of wood and forest social-ecological systems in distinctive basins, this complex unfolding of change described in Indonesia challenges the notion of a social-ecological system making a sudden 'flip' between basins. In the Indonesian example, stewardship forestry institutions have often been accompanied with weakness in other institutions that deliver rule of law, transparency and effective policing. This leads to continuance of patterns of forest use that are consistent with the forest exploitation basin – expressed as illegal frontier logging and forest clearing, even when stewardship forestry is notionally established. Further, this transition between forest exploitation and stewardship forestry is also confounded by the simultaneous pressure of late global industrialisation drivers pushing the transition of stewardship forest to wood cultivation and forest ecosystem management, even before the transition to stewardship forestry has been completed and had a chance to stabilise. In developed countries the stewardship forestry basin was in place for long decades and built resilience through development that was nested in mutually supportive social institutions (such as effective rule of law, adequate and robust policing and governmental transparency). The Indonesian case suggests that changes in wood and forest social-ecological system basins can occur more as messy progression, rather than as the 'flip' often suggested in resilience thinking. It highlights how the larger panarchy influences of modernity and industrialisation, as global phenomena, with their own timetables, impact on individual wood and forest social-ecological systems in different ways. Such timetable differences are important to keep in mind when generalised developmental explanations of resource use social-ecological system evolution (including the one used here) are used.

Acknowledgements

Starting after the heading, *Introduction*, this chapter is a revised and updated version of the original paper; Russell Warman, *Decentralization and forestry in the Indonesian Archipelago: beyond the Big Bang*, South East Asia Research (24, 1) pp. 23–40 Copyright © 2016 (SAGE Publications). Reprinted by permission of SAGE Publications.

Notes

1 See, for example Galudra and Sirait (2009) for a description of the use of hydrological concerns to justify central control of forests in colonial Java.

2 'Sphere of interest' is used in this context to describe a geographic range of interest based on the spatial sphere of activity that will affect actors' interests.

3 *Reformasi* is used here to refer to the period of governance reforms undertaken after the end of the Soeharto government in 1998.
4 143 million hectares was 76 per cent of the total 187 million hectares of land in Indonesia. For details on extent of forest and tenure see Ministry of Forestry (2009) note 8, at p. 9.
5 By 1929 the Javan Forest Service was well established, with nearly 6,000 staff – a level of staffing and staff to forest area ratio much higher than any other comparable jurisdiction in Southeast Asia at the time (Peluso and Vandergeest, 2001).
6 The Outer Islands refer to the islands of Indonesia outside Java, Bali and Madura.
7 For a discussion of the ideas promulgated in the 1960s for the use of forests to advance economic development in the non-western world see Westoby (1987).
8 'Cutting during the flood', a method of hand logging after which this period was named. The method utilised rivers to float logs to markets and involved many small scale local operators (Ross, 2001).
9 Source: FAOSTAT (2014), wood panel logs conversion factor from the Contreras-Hermosilla *et al.* (2007), and plantation wood compiled from various Indonesian Ministry of Forestry reports.
10 *Otda* from *Otonomi Daerah*, meaning Regional Autonomy (see Arnold, 2008).
11 The law granted the Ministry of Forestry responsibility for 120 million hectares of forest (Irawan *et al.*, 2013).
12 Fukuoka uses the term to generically describe rule by a non-ideologically driven ruler using a personal following to wield power, rather than the traditional Indonesian association with historic sultans and their re-emergence as part of *adat* revival. See Van Klinken (2007) for discussion on the latter in particular, and its relationship to the concept's utilisation by Fukuoka.
13 *Inisiatif Satu Peta* or the One Map Initiative was formalised under Law No 4/2011 on Geospatial Information, Information Geospatial Agency. It is intended to provide Indonesia with a single mapping system from which all government agencies will base decisions in order to improve transparency and efficiency (Santosa *et al.*, 2013).

References

Agrawal, A. (2003). Sustainable governance of common-pool resources: context, methods, and politics. *Annual Review of Anthropology*, 32, 243–262.

Agrawal, A., Chhatre, A. and Hardin, R. (2008). Changing governance of the world's forests. *Science*, 320, 1460–1462.

Ahmad, M.S. and Talib, N.B.A. (2011). Decentralization and participatory rural development: a literature review. *Contemporary Economics*, 5, 58–67.

Arnold, L.L. (2008). Deforestation in decentralised Indonesia: What's law got to do with it. *Law Environment and Development Journal*, 4, 75–100.

Astuti, R. and Mcgregor, A. (2017). Indigenous land claims or green grabs? Inclusions and exclusions within forest carbon politics in Indonesia. *The Journal of Peasant Studies*, 44, 445–466.

Barr, C., Resosudarmo, I.A.P., Dermawan, A. and McCarthy, J. (eds.) (2006a). *Decentralization of forest administration in Indonesia: implications for forest sustainability, economic development, and community livelihoods*. Bogor, Indonesia: Center for International Forestry Research.

Barr, C., Resosudarmo, I.A.P., Dermawan, A. and McCarthy, J. (2006b). Forest administration and forestry sector development prior to 1998. In: Barr, C., Resosudarmo, I.A.P., Dermawan, A. and McCarthy, J. (eds.). *Decentralization of forest administration in Indonesia: implications for forest sustainability, economic development, and community livelihoods*. Bogor, Indonesia: Center for International Forestry Research.

Barr, C., Resosudarmo, I.A.P., Dermawan, A., Mccarthy, J. and Moeliono, M. (2006c). Decentralization and recentralization in Indonesia's forestry sector: summary and recommendations. In: Barr, C., Resosudarmo, I.A.P., Dermawan, A. and McCarthy, J. (eds.). *Decentralization of forest administration in Indonesia: implications for forest sustainability, economic development and community livelihoods*. Bogor, Indonesia: Center for International Forestry Research.

Barr, C., Resosudarmo, I.A.P., Dermawan, A. and Setiono, B. (2006d). Decentralization's effects on forest concessions and timber production. In: Barr, C., Resosudarmo, I.A.P., Dermawan, A. and McCarthy, J. (eds.) *Decentralization of forest administration in Indonesia: implications for forest sustainability, economic development and community livelihoods*. Bogor, Indonesia: Center for International Forestry Research.

Barr, C., Resosudarmo, I.A.P., McCarthy, J. and Dermawan, A. (2006e). Forests and decentralization in Indonesia: an overview. In: Barr, C., Resosudarmo, I.A.P., Dermawan, A. and McCarthy, J. (eds.). *Decentralization of forest administration in Indonesia: implications for forest sustainability, economic development and community livelihoods*. Bogor, Indonesia: Center for International Forestry Research.

Blaser, J., Kuchli, C., Colfer, C.J.P. and Capistrano, D. (2005). Introduction: background to a global exchange. In: Colfer, C. and Capistrano, D. (eds.). *The politics of decentralization. Forests, people and power*. London: Earthscan.

Brightman, R.A. (1987). Conservation and resource depletion: the case of the boreal forest Algonquians. In: McCay, B.J. and Acheson, J.M. (eds.). *The question of the commons: the culture and ecology of communal resources*. Tucson, AZ: University of Arizona Press.

Bullinger, C. and Haug, M. (2012). In and out of the forest: decentralisation and recentralisation of forest governance in East Kalimantan, Indonesia. *Austrian Journal of South-East Asian Studies/Österreichische Zeitschrift für Südostasienwissenschaften*, 5, 243–262.

Burgess, R., Hansen, M., Olken, B.A., Potapov, P. and Sieber, S. (2012). The political economy of deforestation in the tropics. *The Quarterly Journal of Economics*, 127, 1707–1754.

Colfer, C.. and Resosudarmo, I.A.P., (eds.) (2002). *Which way forward. People, Forest, and Policy Making in Indonesia*. Resources for the Future, Washington DC.

Contreras-Hermosilla, A., Doornbosch, R. and Lodge, M. (2007). The economics of illegal logging and associated trade. Paris: Organisation for Economic Co-operation and Development (OECD).

Conyers, D. (1983). Decentralization: the latest fashion in development administration? *Public Administration and Development*, 3, 97–109.

Crouch, H. A. (2010). *Political reform in Indonesia after Soeharto*. Singapore, Institute of Southeast Asian Studies.

Davidson, J. and Henley, D. (eds.) (2007). *The revival of tradition in Indonesian politics: the deployment of adat from colonialism to indigenism*. Oxon: Routledge.

Dietz, T., Ostrom, E. and Stern, P.C. (2003). The struggle to govern the commons. *Science*, *302*, 1907–1912.

FAOSTAT (2014). *Statistical databases*. Rome: Food and Agriculture Organization of the United Nations.

Ferguson, I. and Chardasekharan, C. (2005). Paths and pitfalls of decentralization for Sustainable Forest Management: experiences of the Asia Pacific region. In: Colfer, C. and Capistrano, D. (eds.). *The politics of decentralization*. London: Earthscan.

Fukuoka, Y. (2013). Oligarchy and democracy in post-Suharto Indonesia. *Political Studies Review*, 11, 52–64.

Galudra, G. and Sirait, M. (2009). A discourse on Dutch colonial forest policy and science in Indonesia at the beginning of the 20th century. *International Forestry Review*, 11, 524–533.

Gregersen, H. M., Contreras-Hermosilla, A., White, A. and Phillips, L. (2005). Forest governance in federal systems: an overview of experiences and implications for decentralization. In: Colfer, C. and Capistrano, D. (eds.). *The politics of decentralization. Forests, people and power*. London: Earthscan.

Hadiz, V.R. (2010). *Localising power in post-authoritarian Indonesia: a Southeast Asia perspective*. Stanford, CA: Stanford University Press.

Hadiz, V.R. and Robison, R. (2013). The political economy of oligarchy and the reorganization of power in Indonesia. *Indonesia*, 96, 35–57.

Hardin, G. (1968). The tragedy of the commons. *Science*, 162, 1243–1248.

Hyde, W.F. (2012). *The global economics of forestry*. New York: Routledge.

Irawan, S., Tacconi, L. and Ring, I. (2013). Stakeholders' incentives for land-use change and REDD+: The case of Indonesia. *Ecological Economics*, 87, 75–83.

Johnson, C. (2013). *Indonesia: forest rights of indigenous peoples affirmed* [Online]. Washington DC: Law Library of Congress. www.loc.gov/lawweb/servlet/lloc_news?disp3_l205403604_text.

Karsenty, A. and Ongolo, S. (2012). Can "fragile states" decide to reduce their deforestation? The inappropriate use of the theory of incentives with respect to the REDD mechanism. *Forest Policy and Economics*, 18, 38–45.

Kartasubrata, J. (1985). Forest policy, legislation and administration in Indonesia. *Nederlands Bosbouwtijdschrift (Netherlands)*, 57, 167–175.

Kimura, E. (2010). Proliferating provinces: territorial politics in post-Suharto Indonesia. *South East Asia Research*, 18, 415–449.

Larson, A. M. (2005). Democratic decentralization in the forestry sector: lessons learned from Africa, Asia and Latin America. In: Colfer, C. and Capistrano, D. (eds.). *The politics of decentralization: forests, power and people*. London: Earthscan.

Malley, M. (2003). New rules, old structures, and the limits of Indonesia's democratic decentralization. In: Aspinal, E. and Fealy, G. (Eds.). *Local power and politics in Indonesia: decentralisation and democratisation*. Singapore: Institute of Southeast Asian Studies.

McCarthy, J. (2000). The changing regime: forest property and *reformasi* in Indonesia. *Development and Change*, 31, 91–129.

McCarthy, J. (2007). Sold down river: renegotiating public power over nature in Central Kalimantan. In: Nordholt, H.S. and Van Klinken, G. (eds.). *Renegotiating boundaries: local politics in post-Suharto Indonesia*. Leiden, The Netherlands: KITLV Press.

Ministry of Forestry (2009). Indonesia forestry outlook study. In: FAO (ed.) *Working Paper Series*. Bangkok: Food and Agricultural Organisation of the United Nations.

Moeliono, M., Wollenberg, E. and Limberg, G. (2009). *The decentralization of forest governance: politics, economics and the fight for control of forests in Indonesian Borneo*. London: Earthscan.

Murdiyarso, D., Dewi, S., Lawrence, D. and Seymour, F. (2011). *Indonesia's forest moratorium: a stepping stone to better forest governance?* Bogor, Indonesia: Center for International Forestry Research.

Myers, R., Intarini, D., Sirait, M.T. and Maryudi, A. (2017). Claiming the forest: Inclusions and exclusions under Indonesia's 'new' forest policies on customary forests. *Land Use Policy*, 66, 205–213.

Nordholt, H.S. and Van Klinken, G. (2007). *Renegotiating boundaries: local politics in post-Suharto Indonesia.* Leiden, The Netherlands: KITLV Press.

Obidzinski, K. and Chaudhury, M. (2009). Transition to timber plantation based forestry in Indonesia: towards a feasible new policy. *International Forestry Review*, 11, 79–87.

Obidzinski, K. and Dermawan, A. (2012). Pulp industry and environment in Indonesia: is there sustainable future? *Regional Environmental Change*, 12, 961–966.

Olson, M. (1965). *The logic of collective action: public goods and the theory of groups.* Cambridge, MA: Harvard University Press.

Ostrom, E. (1990). *Governing the commons: the evolution of institutions for collective action.* Cambridge: Cambridge University Press.

Palmer, C. and Engel, S. (2007). For better or for worse? Local impacts of the decentralization of Indonesia's forest sector. *World Development*, 35, 2131–2149.

Pandey, D. and Brown, C. (2000). Teak: a global overview. *Unasylva*, 51, 3–13.

Peluso, N. (1992). *Rich forests, poor people: forest access and control in Java.* Berkeley, CA: University of California Press.

Peluso, N.L. (1995). Whose woods are these? Counter-mapping forest territories in Kalimantan, Indonesia. *Antipode*, 27, 383–406.

Peluso, N.L., Afiff, S. and Rachman, N.F. (2008). Claiming the grounds for reform: agrarian and environmental movements in Indonesia. *Journal of Agrarian Change*, 8, 377–407.

Peluso, N.L. and Vandergeest, P. (2001). Genealogies of the political forest and customary rights in Indonesia, Malaysia, and Thailand. *Journal of Asian Studies*, 60, 761–812.

Radjawali, I. and Pye, O. (2015). *Counter-mapping land grabs with community drones in Indonesia. Proceedings of the Land Grabbing, Conflict and Agrarian – Environmental Transformations: Perspectives from East and Southeast Asia, Chiang Mai, Thailand.* Chang Mai, Thailand: BRICS Initiatives for Critical Agrarian Studies (BICAS).

Resosudarmo, I.A.P. (2003). Shifting power to the periphery: the impact of decentralisation on forests and forest people. In: Aspinal, E. and Fealy, G. (eds.). *Local power and politics in Indonesia: decentralisation and democratisation.* Singapore: Institute of Southeast Asian Studies.

Ribot, J.C., Agrawal, A. and Larson, A.M. (2006). Recentralizing While Decentralizing: How National Governments Reappropriate Forest Resources. *World Development*, 34, 1864–1886.

Ricklefs, M.C. (2001). *A history of modern Indonesia since c. 1200.* Stanford, CA: Stanford University Press.

Rodríguez-Pose, A. and Ezcurra, R. (2010). Does decentralization matter for regional disparities? A cross-country analysis. *Journal of Economic Geography*, 10, 619–644.

Rodríguez-Pose, A. and Sandall, R. (2008). From identity to the economy: analysing the evolution of the decentralisation discourse. *Environment and Planning C: Government and Policy*, 26, 54–72.

Ross, M.L. (2001). *Timber booms and institutional breakdown in Southeast Asia.* Cambridge: Cambridge University Press.

Royo, N. and Wells, A. (2012). Community based forest management in Indonesia: a review of current practice and regulatory frameworks. Background paper for The Forest Dialogue on Investing in Locally Controlled Forestry. *Yogyakarta, Indonesia*, 6–9.

Sachs, J.D., (2012). From millennium development goals to sustainable development goals. *The Lancet*, 379, 2206–2211.

Santosa, M.A., Khatarina, J. and Suwana, A.S. (2013). The progress on governing REDD+ in Indonesia. *International Journal of Rural Law and Policy*, 1, 1–17.

Sayer, J., Elliott, C., Barrow, E., Gretzinger, S., Maginnis, S., McShane, T. and Shepherd, G. (2005). The implications for biodiversity conservation of decentralized forest resources management. In: Colfer, C. and Capistrano, D. (eds.). *The politics of decentralization: forests, power and people.*

Siswanto, W. and Wardojo, W. (2005). Decentralization of the forestry sector: Indonesia's experience. In: Colfer, C. and Capistrano, D. (eds.). *The politics of decentralization: Forests, power and people.* London: Earthscan.

Van Klinken, G. (2007). Return of the sultans: the communitarian turn in local politics. In: Davidson, J. and Henley, D. (eds.). *The revival of tradition in Indonesian politics: the deployment of adat from colonialism to indigenism.* Oxon: Routledge.

Vlekke, B.H.M. (1959). *Nusantara: a history of Indonesia.* The Hague, The Netherlands: W. van Hoeve Ltd.

Westoby, J. (1987). *The purpose of forests: follies of development.* New York: Basil Blackwell.

Wright, G. (2011). Indigenous people and customary land ownership under domestic REDD+ frameworks: a case study of Indonesia. *Law Environment and Development Journal*, 7, 119–131.

Wunsch, J.S. (2013). Analyzing self-organized local governance initiatives: are there insights for decentralization reforms? *Public Administration and Development*, 33, 221–235.

Yudhoyono, S.B. (2014). *Indonesian president calls on successor to continue moratorium on forest concessions* [Online]. Bogor, Indonesia: Center for International Forestry Research. Available: www.cifor.org/forestsasia/indonesian-president-calls-on-successor-to-continue-moratorium-on-forest-concessions/ [Accessed 27 May 2014].

7 A discursive turn

Foresters' speak, and transitions in New Zealand

Introduction

The third national case study is of the country of New Zealand. New Zealand is in the South Pacific to the east of Australia. A large island nation with a wet temperate climate and rich soils, it has a significant wood industry and unique indigenous forests. Of note, New Zealand is well advanced in the transition to distinctive basins of wood cultivation and forest ecosystem management. The country has a large and growing wood products industry based on the cultivation of exotic tree species in industrial plantations. Meanwhile its natural forest ecosystems are managed for non-wood values. The case study here takes the third 'tack' of the discursive and analyses changes in wood and forest social-ecological systems using a blend of discourse and quantitative text (content) analysis. It seeks understanding and insight into the changes between basins of the wood and forest social-ecological system in the discussions and language used by foresters in New Zealand during this period.

The main texts used are the corpus of the journals of the national professional forester's institute over the 87 year period from 1927 until 2013.[1] When the journal commenced, New Zealand was fast approaching the limits of its natural forests' abilities to meet wood demand and was developing stewardship forestry institutions in response (Fleet, 1984; Roche, 1987). The journal was an expression of the emergence of the stewardship forestry basin; of the establishment of formal institutions intended to bring about sustainable wood production in New Zealand's forests. Over the subsequent decades, forestry in New Zealand transited through a period of stewardship forestry in its remaining natural forests to intensive wood cultivation. The shifts in wood production from natural forest exploitation to cultivated wood sources are shown in Figure 1.3. In 2011 New Zealand had 2,090,000 hectares of exotic tree plantations and 8,063,000 hectares of indigenous forest ecosystems (FAO, 2014). Plantation sourced wood has gone from being close to 0 per cent of wood supply in the 1920s to almost 100 per cent now. In addition, New Zealand now produces volumes of plantation wood 20 times greater than its natural forests were producing at their peak 60 years ago (Figure 1.3). A near complete transition of wood production from natural forest exploitation to wood cultivation has occurred. Given this, the writings in the journal are a potentially rich data source for understanding the transitions between basins in the wood and forest social-ecological system of the country. This analysis sought discursive reportage of changes in the basins and also of the resilience of basins and resistance to change. It looks for evidence of discussion that implied recognition of fixed states (stable basins) and stressors that were forcing potential change on those systems. Finally, it considers how tensions over the pursuit of changing technologies and shifts in social values were reconciled with forestry's core aims of sustainability and creating stable systems in equilibrium.

Discursive analysis approaches text as the site of social actions (Wetherell, 2001). In particular, it looks at the social actions of key actors, specifically foresters, in the wood and forest social-ecological system. Wetherell (2001) provides three useful approaches to thinking about discourse analysis as a way to understand social phenomena. First, she refers to *correlation* – noting language that is a reflection or response to social circumstances. Second, *choice* – the illumination of a writer's choice of words and their intention to convey selected meaning. This approach seeks to understand the influence of the writer (speaker) on their social environs, not merely its correlation with the social. Finally, *critique* – this is where the analyst, following the logic of *choice*, asks of the discourse what choices did the writer make, and perhaps most usefully, why? Wetherell observes that power relations in particular can be studied in discourse analysis and Van Dijk (2001) also stresses the importance of addressing power relations in a critical discourse analysis. To some extent, the role of power relationships between different interest groups in the stewardship forestry basin was addressed in the previous case study of Indonesia. This chapter seeks to deepen this by focusing on the central role of forestry professionals; how do they confront and influence change in the wood and forest social-ecological system?

There is limited literature using discourse and content analysis related to forestry. Findings tend to focus on conflict and change in forestry systems and how key groups have deployed language to affect (or deflect) change (for example, Bell, 1995; Hillier, 2003; Wolf and Klein, 2007), although Xu and Bengston (1997) take a correlation approach, analysing the language for markers of change in forestry in the United States. The limited body of literature on such reflexive analysis in forestry possibly arises from a combination of the pragmatic nature of the forestry endeavour and a concomitant acceptance of the rationale for stewardship forestry as self-evident. And yet, the circumstances in which a social institution might arise are subject to all manner of change. Thus, there is considerable value in an enquiry such as found in this chapter, especially as it facilitates reflection on these seemingly self-evident truths.

Methods

The *New Zealand Journal of Forestry* contains a mix of content, including peer reviewed articles, letters to the editor and editorials. These traverse a period of significant changes in forestry in New Zealand. All editions of the journal from 1925 to 2013 are freely available on the New Zealand Institute of Forestry website: www.nzjf.org.nz/. To limit the corpus size and to keep the analysis manageable, but giving effective time coverage, just one edition from each year was selected. Until the mid-60s there was only one edition. From 1964 the journal started producing two issues a year. The first issue aimed to 'contain as wide a variety of papers on forestry as we can secure', while the second was focused on papers from the Institute's annual symposium (Jackson, 1964, p. 6). Based on this, the first edition for each year was selected as it indicated a wider range of themes and a lower risk of special themes influencing the results. From 1986 four issues were published. The editorial at the time spoke of seeking a wide range of articles with no particular issue focused on special content or subjects (Mead, 1986). Therefore, in keeping with the previous period, issue number 1 from each year was chosen. All up, 87 journals, one from each year of an 89 year period, were selected for the corpus (the journal was not published in 1942 and 1943, presumably due to the interruption of the Second World War).

The downloaded issue files were cleaned of duplicate pages and text, such as headings and page numbers that might have biased quantitative text analysis[2] through a mix of manual and automated processes. To smooth out potential noise from single issues or even articles within an issue, the corpus was partitioned into 18 five year blocks (note that 1940 to 1944 only included 1940, 1941 and 1944, and 2010 to 2014 excluded 2014). The corpus was analysed using a specificity test. Specificity is a statistical measure of the frequency of a particular word's occurrence within a specific period (see Drouin, 2004). It is based on changes in the portion of the total words of that period accounted for by a specific word – thus cancelling out the effect of total words varying across periods. It signifies how much the usage of that word in a specific period (five years in this case) differs from the average across the total corpus duration (89 years in this case). In addition, two other functions were used: *concordance* (a display of words to the left and right of a specified word that show context) and *co-occurrence* (the frequency and average distance [measured in number of words] one word occurs with or near another).

The corpus analysed contained 4,063,681 words[3] with 52,817 distinctive words. Of these, the 49,277 that had less than 100 occurrences of the word in the corpus were removed in an initial shortlisting. From the remaining results, dates, numbers and symbols (for example '%') were also removed. Of the remainder, a list of about 300 words considered most relevant to change in forestry systems (selected based on a course coding of the text) was shortlisted to analyse closely. The specificity test described above was applied to these words. The next analytical approach was to look for change in usage of certain words as indicated by their specificity results. These words were then further analysed using concordance and co-occurrence results to seek out context and surrounding discourse.

Results and analysis

The graphs on the following pages (Figure 7.1) show the specificity results of selected words. To better show the detail of lower specificity scores the graphs have limited the *y*-axis to between 10 and –10. Specificities between 2.0 and –2.0 are not considered statistically significant and are shown by the central greyed range in graphs.

This section has been structured starting with *forestry*, the word denoting the central institutions of the stewardship forestry basin. It then looks at *sustainability*, a term that describes the guiding stewardship forestry logic but might also indicate the emergence of the forest ecosystem management basin. This in turn leads to an exploration of two main tensions acting on the stewardship forestry basin – increasing demand for non-wood values and technological and economic drivers towards increased efficiency, production and intensification, and key words relating to these tensions.

The word *forestry* had a positive specificity from the 1980s onward despite being the central idea of the journal across the period – that is *forestry* made up a greater portion of total words in the five year periods from the 1980s onward than it did in earlier five year periods of the corpus. Why was *forestry* written much more often at the end of the period? One possibility is a change in audience. If the journal was originally written by foresters for foresters then, maybe, things being *forestry* was assumed to be self-evident. A discourse could occur within forestry that has no need to refer to itself if the writer is confident that both they and their reader exist within forestry. If, though, the audience became more diverse later in the journal's publication period, there might be a greater need to declare matters as *forestry*. Another possibility is that this change could indicate an emerging need to defend the conceptual territory of *forestry* to others. It might, for example, be that the profession felt a growing need to press its case, '[t]he imperative need for the forestry sector to … publish its findings is now clear' (Holloway, 1983, p. 6), including to governments where, 'there is a hands-off attitude by the government to production forestry which is mirrored in the government departments' (Goulding, 2013, p. 44). Or, it may stem from a need to defend or explain itself to an increasingly critical public: 'there is an urgent need to challenge much of the negative comment about forestry contribution and its future potential', said Wink Sutton, the new president of the Institute (quoted in Anon, 1988, p. 26).

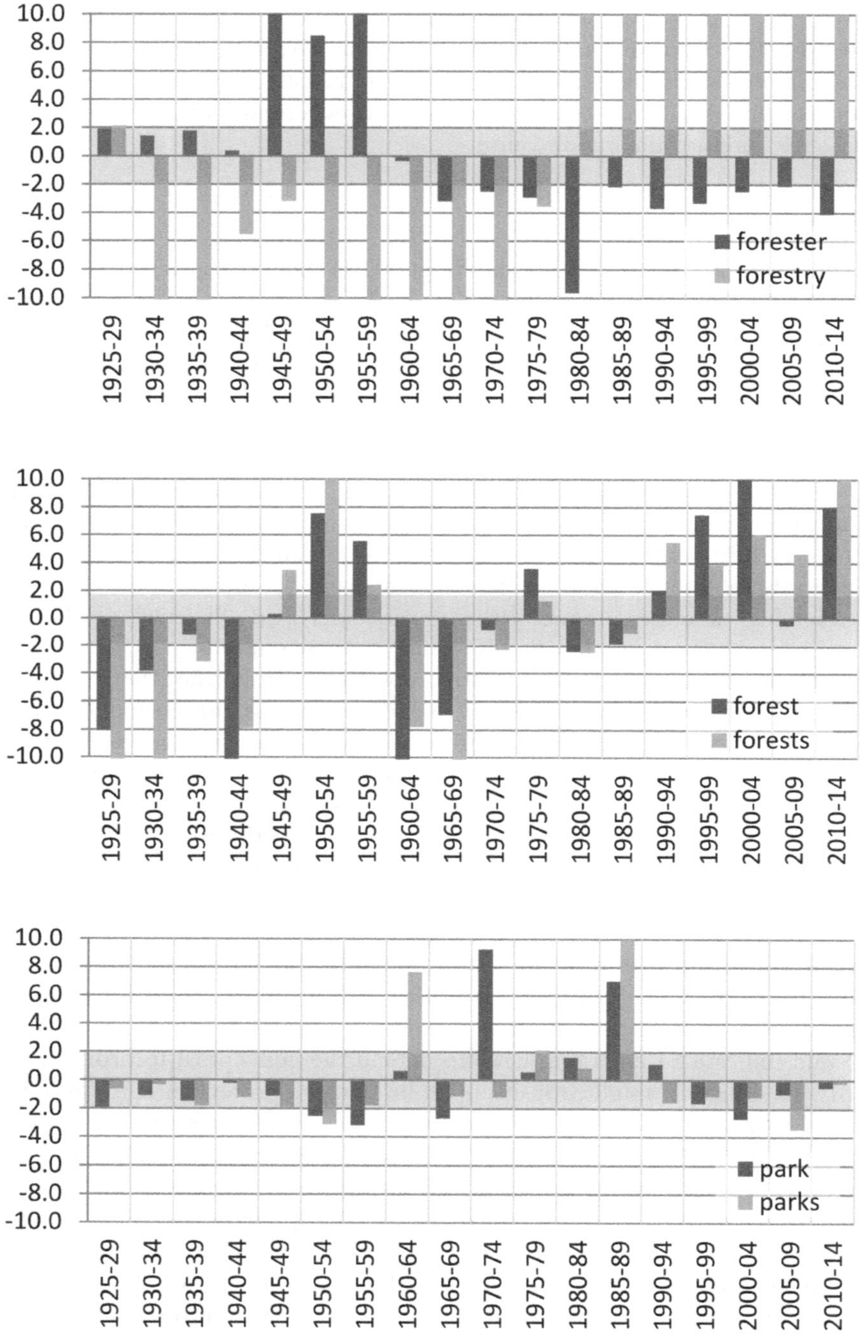

Figure 7.1 Specificity results of selected words noted in the following discussion.

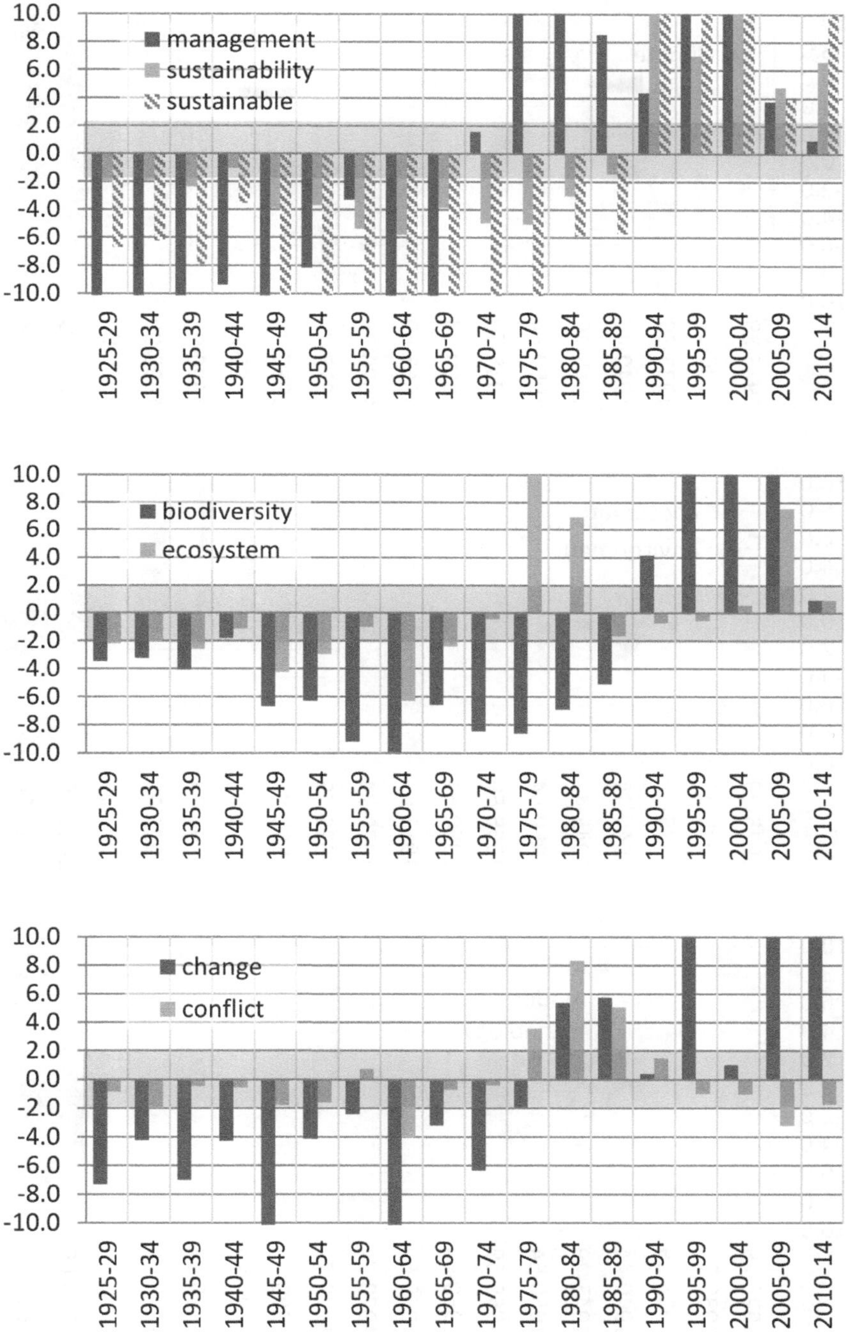

Figure 7.1 Continued.

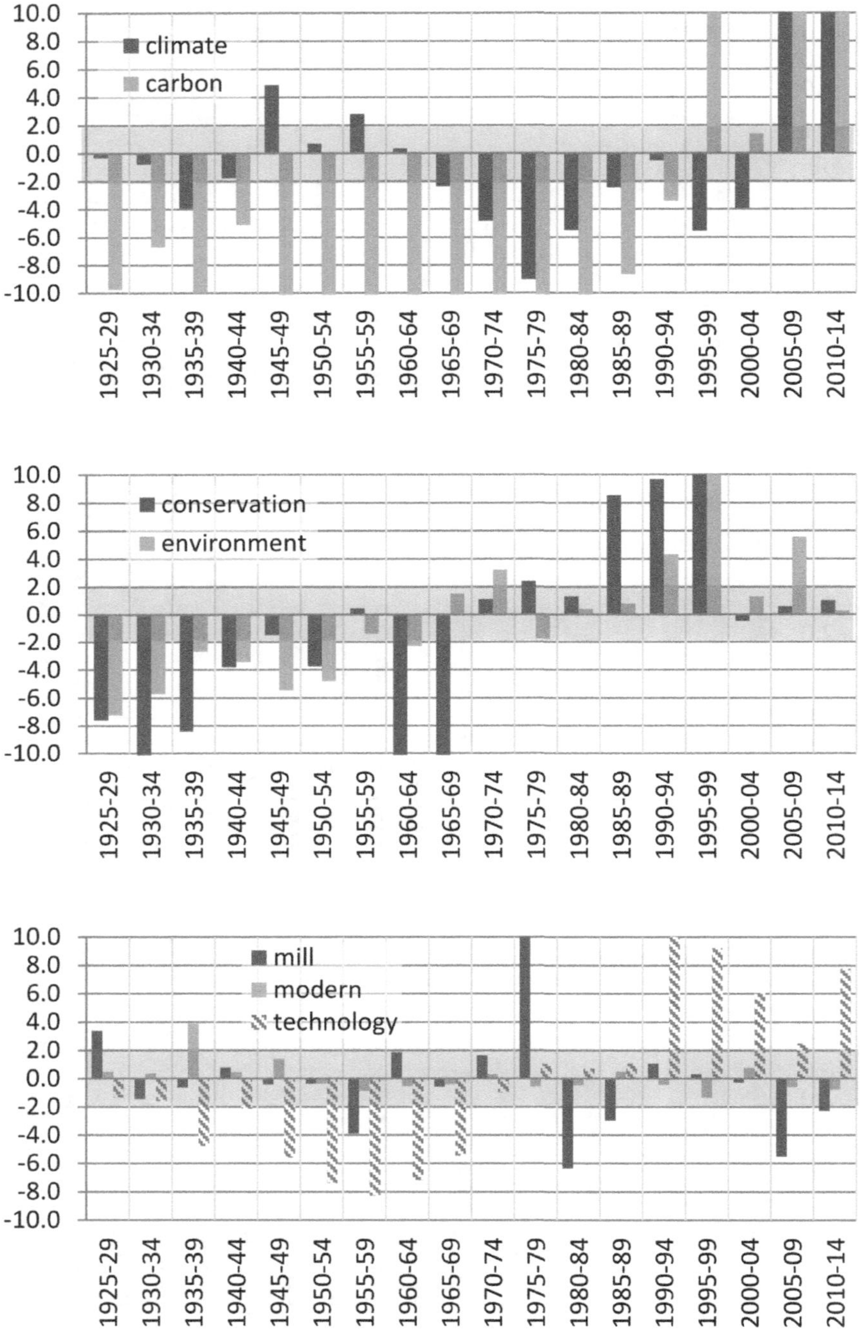

Figure 7.1 Continued.

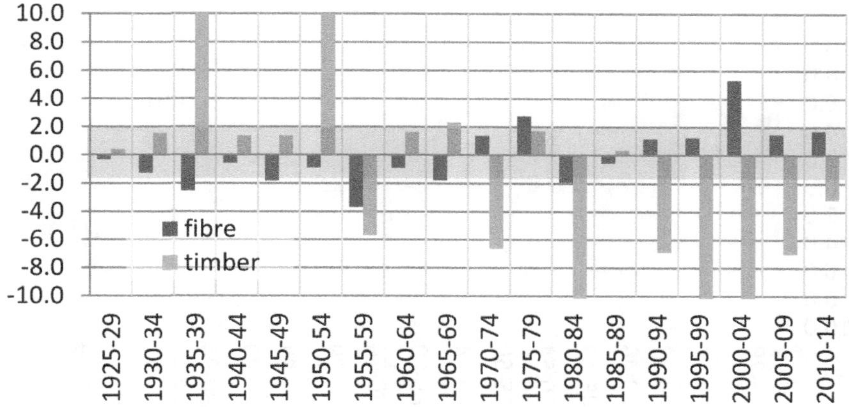

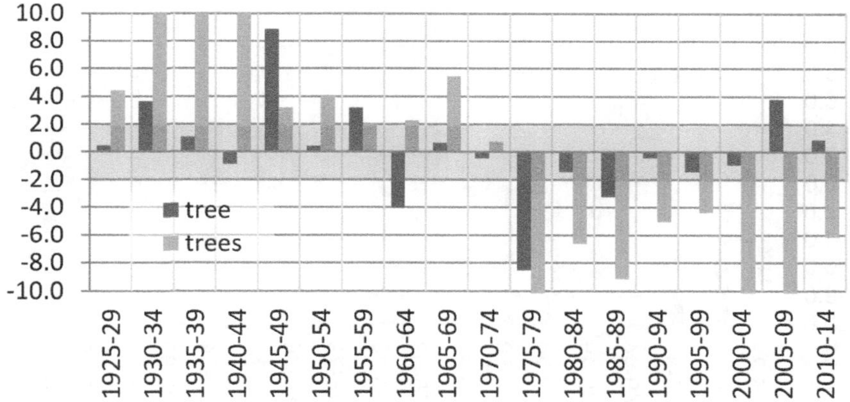

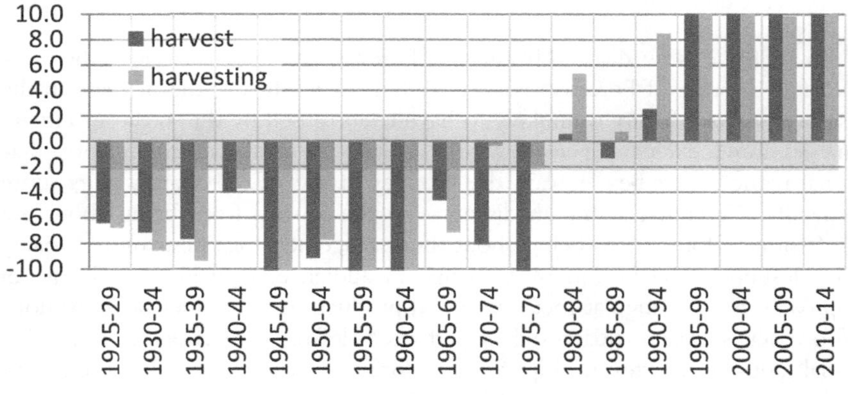

Figure 7.1 Continued.

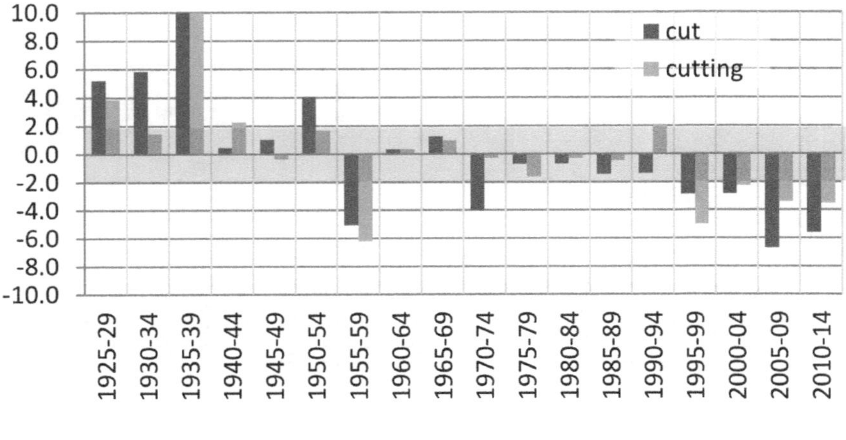

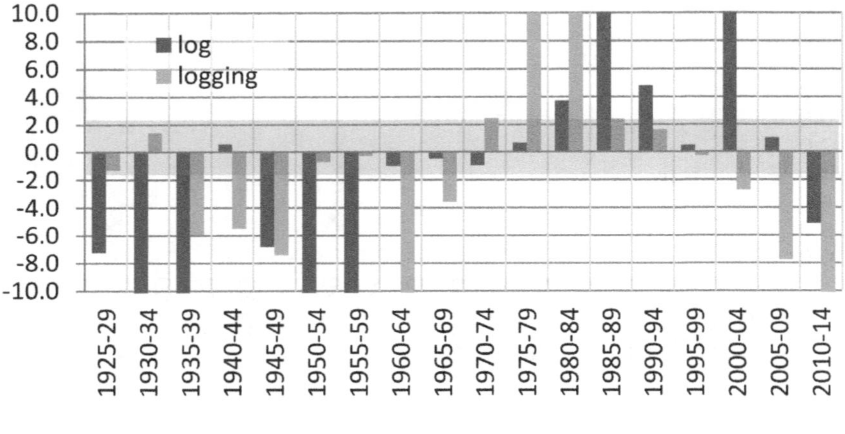

Figure 7.1 Continued.

In 1988 the New Zealand Institute of Foresters changed its name to the New Zealand Institute of Forestry (Halkett, 1989). Nevertheless, the increase in the use of the word *forestry* is unlikely to have been caused by this as the specificity analysis above excluded *Forestry* with a capital *F*. But, the change in organisational name might best explain the changes in use of the word *forestry* more generally. Certainly, a shift in discourse in the journal from being about the profession of foresters to being about the management of forests more broadly would support the notion of a change in audience, as well as a change in approach, from a singular focus on wood production to a wider range of non-wood values. Again, this could in part be a defensive reaction to a growing social critique of stewardship forestry forest management. The President's address in 1982 (Rockwell, 1982, p. 23) notes the Institute's mission was 'considerably widened' from the historic 'further the development of forestry' and 'interest of the profession', to promoting 'the best use of New Zealand's resources', encouraging 'wise use of forests and forest land', as well as supporting the profession. This backs the idea that there was a widening of concern beyond the profession to a wider audience. The shift in emphasis from *forester* to *forestry* in the name of the Institute attempts to embrace this and reflects the greater range of values expected from forests. It describes the push to go beyond wood extraction as the basic stewardship forestry activity.

As has been noted, sustainability of wood supply was the motivating consideration of the stewardship forestry basin. It was expected then that the words *sustainable* and *sustainability* might both feature in early discussions, nevertheless both showed a pattern of positive specificity in the period from 1986 onwards (and negative before). *Perpetuity* is another word that might be associated with these ideas but, again, it first appears twice in the 1940s before becoming prominent thereafter. Like *sustainable*, it is not correlated with the emergence of the stewardship forestry basin. *Sustained*, another word that conveys this ongoingness, does occur almost as often as *sustainable*, but its use across the corpus is more widespread with a pattern of lower use in the earliest and latest years only. An obituary for Leon Macintosh Ellis (Anon, 1944, p. 7) identifies, as one of his main achievements as first Director of Forestry, the legislation that secured 'the basic land requirement of all forestry – permanency of use and tenure'. It is clear that sustainability, as it is now considered, was the goal but the term itself came into use only as part of increasing demands for ecological forest ecosystem management – a source of tension and challenge to the wood supply logic of the stewardship forestry basin.

Specifically, *sustainability* and *sustainable* become widely used from the end of the 1980s. The rise of the use of *sustainable* is consistent with its application in the context of a broader global rise in the social awareness of environmental issues (Dunlap and York, 2008). The word *sustainability* is then associated with the rise of the concept of forests as ecosystems that deliver a diverse range of services and values, rather than just as a perpetual source of wood. This is likely to be associated with the rise in sustainability thinking that emerged globally after the events such the release of the Bruntland Commission's 1987 report *Our Common Future* and the Rio Earth Summit of 1992 (Mebratu, 1998). In this sense, while forestry had clearly pursued what is now considered *sustainability* in wood production, it embraced the term only in response to the emergence of sustainability within the global environmental discourse. Again, it is also possible that this reflects a need to defend the work of forestry in an environment where it was now often under attack precisely because of perceived lack of sustainability. In this sense, foresters, a profession established to create sustainability (of forests for reliable wood supply), were now being questioned as to the credentials of their sustainability.

Correlated with the rise of the term *sustainable* was usage of the term *management*, which increased from the 1970s until an apparent tapering toward the end of the period researched. Key co-occurrent words for *management* included *forest (forests), sustainable, practices, indigenous* and *planning (plan)*. The concordance (within 11 words) of *management* and *sustainable* is 191 times, but only from 1991 onwards. The term *sustainable forest management* occurs 55 times from 1993 onward (never before). In most cases the term is followed by words denoting its bureaucratic integration; for example, *plan, rules, criteria, standards* and *framework*. Reflecting the rise of sustainability in the development discourse noted above, the 1993 Forest Amendment Act in New Zealand included requirements for sustainable forest management plans. So, for example, '[t]he Act ... permits the export of rimu and beech sawn timber, provided it is harvested under a registered sustainable forest management plan, or a sustainable forest management permit' (Baddeley, 1993). This broader shift in use of the term *management* may also reflect a shift in the nature of forestry to one that is more managerial. This could indicate an increasing managerial turn in western society generally over the same period that would be indicative of such panarchy processes as globalisation, industrialisation and technological innovation affecting change within the wood and forest social-ecological system.

Park and *parks* have a high level of specificity in the 1960s to 1990s, but most noticeably in the period 1985 to 1998. The term *conflict* also surges in usage in the late 1970s and 1980s. This was a period when the political debate about natural forest logging was leading up to its eventual cessation around 1990. However, the debate between ideas of protecting natural forest from logging and managing it for timber production had been occurring for some time. The journal editor (Anon, 1946, p. 173) writes of the controversy over Waipoua State Forest:

> [t]he article notes a school of thought born of the sentimental reaction to the pioneering phase of forest destruction which in its extreme form regards the preservation of all indigenous forest in its existing condition ... as a desirable objective in all circumstances. These zealots of "preservation by non-use" ... hold their beliefs with irrational and unyielding depth of conviction.

While then acknowledging some legitimacy to these demands, the editor went on to caution that these claims 'must be kept in proportion if all attempts at silvicultural management of our indigenous forests are not to be frustrated'. In 1976 Leslie (1976, p. 6) noted that:

> [i]t is not so long ago that forestry was in the forefront of the conservation movement. Now it has become one of the prime targets of the movement. Somewhere over the last two decades forestry and conservation parted company.

The change to emphasis on forest ecosystem management can be clearly seen in the upswing in usage of the terms *biodiversity, conservation* and *environment* from the 1980s onwards. These concerns were reflected in discussions of the time:

> at no other time in our history has the value of forests been so highly recognised. Nevertheless, governments world-wide are struggling to come up with the right set of tools to value forests for their carbon sequestration, clean water, human environment and biological diversity.
>
> (Charteris *et al.*, 2010, p. 5)

The increased usage of these terms is likely to reflect the growth in environmental awareness already noted. It shows the responsiveness of forestry discourse to the general increase in societal concern for these issues. Notably, the rise of concern about climate change and the role of forests in mitigation and adaptation are reflected by changes in key words over time. Indeed, the term *climate change* occurs 182 times, with only four occurrences before 1990 and then only in relation to prehistoric and historic climate change. The rise of the use of the word *carbon* is another reflection of the rise of interest in climate change. In the last few decades their language suggests it has become a significant concern for foresters in New Zealand.

The term *change* was used more than twice as often from 1980 onward – and that, after its use as part of the phrase *climate change* is discounted. So, even separating out concern about climate change specifically, foresters in New Zealand wrote *change* more often in this period. The first part of this increased use of *change* coincides with the period in which there was considerable public discussion about the type of wood and forest social-ecological system the country should have. As noted above this was when foresters, a profession founded on the very idea of sustainability, were confronted with charges of unsustainable practice. In the 1980s there was considerable discussion about the cessation of logging within public natural forests – an end to stewardship forestry over all remaining natural forest. And, towards the end of the decade the decision was taken (not as one decisive policy change but through a series of policy developments over several years) for New Zealand to make a transformative state change – out of the stewardship forestry basin to one of forest ecosystem management and a geographically distinct wood cultivation basin.

Another pressure operating to change basins in wood and forest social-ecological systems is technology change. This change was exemplified by the organisational realignment of the key national forestry research body, as noted on their website:

> [t]he New Zealand Forest Research Institute Ltd (Scion) is … focused on improving the international competitiveness of the New Zealand forestry industry and building a stronger bio-based economy. Scion has evolved from a company with a focus on the delivery of research, technology and service solutions to the forest and forest products sector to a company focusing on the entire field of plant biomaterials. The aim of Scion is to

> improve the competitiveness of New Zealand's forestry industry and advance the use of biomaterials.
>
> (Treasury, 2015)

This description makes pointed reference to what it has been in the past. There is a need to explain the organisation's past. But why? Is it apology, self-consciousness or does it aim to ensure the barely recognisable new language and mission is understood by those who identify with the 'forest and forest products sector'? Awareness of change and the transition is explicitly communicated. Scion's description talks of what it was, historically, and what it has 'evolved' into. It was an organisation that provided research and development services to the 'forest and forest products sector'. No longer is the talk of forest products – *timber* and *wood* – but of *biomaterials* and *fibre*. The statement indicates a significant evolution in how forestry and wood production is approached in New Zealand. This is reflective of changes in the technology of wood processing and societal and market demands. What is also striking here is the ability to discuss the industry in a way that is devoid of mention of sustainability and non-wood values. It suggests that when *forestry* is working on the agronomic cultivation of wood fibre, in landscapes entirely distinct from indigenous forests, it is able to more fully focus on the task of wood fibre and product production. It is free of the need to manage forest ecosystems for non-forest values found in natural forest management.

One of the significant shifts in wood processing technology globally is the trend to composite wood products – the breaking down of wood into smaller parts and reconstituting them into the required product. This breaking down is done by sawing, peeling, flaking, chipping and pulping. Wood becomes fibre. Within the corpus the use of the word *fibre* increased slightly over the period. Part of this could be a shift of usage. Where once *wood* would have been used to describe the central product of forestry, *fibre* is increasingly the preferred term, such as in:

> [t]here is also a move away from growing longer rotation, large trees by traditional, intensive management towards lower-cost, short rotation management for production of fibre. As such those countries which can produce short rotation wood and fibre will hold a competitive advantage.
>
> (Wijewardana, 2005, p. 38)

It is also possible that the increasing use of *fibre* reflects the growing role of the pulp and paper sector within the wood products industry. As well as the use of *fibre* for increasingly sophisticated composite wood products, it is also more and more applied within the context of bio-products, as noted in the shift of emphasis undertaken by Scion.

Changes in the use of the word *technology* were observed, with it being used more in the latter part of the period, specifically from the 1990s onwards. There are 31 instances of *new* and *technology* appearing within 11 words, mostly as *new technology*, though only from 1991 onwards, such as in '[g]lobalisation and new technologies have helped the planted forest industry to respond ... using new growing and processing technologies and producing new products' (Wijewardana, 2005, p. 38). It is recognised that technology is important to forestry and changing forestry. There is acknowledgement of new technologies in wood processing and discussion about the application of new technologies to forests practices such as in forest metrics (for example aerial photography, remote sensing, GIS and drones). Technology is discussed as an opportunity and something to embrace – 'forest growers ... benefit from adopting new technology in growing and processing' (Wijewardana, 2005, p. 39), or, 'the need for new sophisticated harvesting technology developments and a skilled workforce' (Loughlin, 1990, p. 16).

The embrace of the modern in stewardship forestry can be seen in the phrase *technology transfer* that also appears from the late 1980s onwards. As well as reflecting a rise in interest, and recognition of the role of technology in general, this also reflects a confidence in New Zealand in its plantation wood cultivation and a belief that it has something to offer the world. There are interesting parallels with the experience of Germany, where acute wood shortages led to stewardship forestry and sustainable yield management. As other economies around the world confronted the same problem of unsustainable resource use they, in turn, took on German forestry expertise. This process of international skills dissemination, then, can become an element within national forester cultures. Jack Westoby (1970. p. 9), speaking as part of the New Zealand Forest Service's Golden Jubilee, spoke of 'the progressive insertion of New Zealand into the world forest and timber economy', and noted that:

> the education of a forester is incomplete without some international experience. And the real gain from service overseas and from multiplying international contacts lies ... in the intellectual cross-fertilization which enables the forester to think about his own problems in new and different ways and, in particular, to see them in the light of a world perspective.
>
> (Westoby, 1970, p. 12)

Again, emphasising forestry's immersion in modernity and globalisation. However, as might be expected, there are contrasting perspectives where this pursuit of the modern is challenged. Technology is discussed as something that is to be approached with caution – '[t]here will always be risks relating to feasibility of new technologies' (Burdon, 2008, p. 5), and '[o]f course, if we pursue ... any other new technology, it must be done with great care and caution' (Moore, 2001, p. 10). This is consistent with the tensions within forestry between the qualities of tradition and timelessness that come from managing forests as long-term stable systems and managing a social-ecological system that is embedded in a constantly evolving modernity. The former sees tree growing and forest stewardship as an activity rooted in tradition (and therefore pre-modernity). The draw of the traditional speaks to the struggles of foresters and their attempts to protect forests from the ravages of an expanding industrial society and then, again, to protect those same forests from a rising social ecological impulse that would remove forests from the hard-won stewardship of the forester. As the New Zealand Forest Service faced restructure in the mid-1980s it was observed that it 'retained a strong commitment to its traditions and long-term goals and was not receptive to the suggestions for change' (Halkett, 1987, p. 20). However, tradition in the context of modernity within which stewardship forestry arises is problematic, 'for justified tradition is tradition in sham clothing and receives its identity only from the reflexivity of the modern' (Giddens, 1990, p. 38). Foresters have this struggle then, being at once embedded within modernity and its constant strive for change and, at the same time, working against its expressions in seeking wood and forest social-ecological systems that are stable and perpetual and grounded in tradition.

Finally, here the three words *cut, log* and *harvest* are investigated. They illustrate distinctive patterns of usage over 90 years. In their verb forms these words represent the basic wood production act – converting a living tree into wood. The terms *cut, log* and *harvest* and their derivatives display three distinct 'waves' of usage. *Cutting* is used most often until the 1950s, *logging* has a higher level of usage during the 1970s and into the 1980s, and *harvesting* becomes more common in the 1980s, but appears rarely beforehand. What do these terms indicate of distinct phases of wood production?

Cut suggests the finding and exploiting of wood from extant natural forests, as in 'for two sawyers to cut 50,000 feet of timber near Timaru' (Clark, 1926). A feature of this phase of forest use is that wood-getting is often entwined with clearing the land for non-forest uses such as settlement and agriculture. It is not just the tree that is *cut* for wood – the forest is *cut* to clear the land, often wastefully, '[t]here was no provision for any control of the licensee, who could go into any part of the district to which his license referred and cut and destroy timber without check' (Clark, 1926, p. 22). There are numerous references to forest that had been *cut-over.* In 1935 the Director of Forestry reported '[b]oth the remaining virgin stands and the cut-over areas carrying regeneration and advance growth are being brought under forest Working Plans' (Anon, 1935). Further, 'misgivings about the general handling of the indigenous forests' in New Zealand forestry are noted in an editorial review of the Summary Report of the Seventh British Commonwealth Forestry Conference of 1957, which lamented '[t]he rapid rate of cutting, the devastated condition of cut-over forests' (Poole, 1959, p. 7). In addition to these notions of *cut* as a frontier exercise, cases of the word *cut* in the earlier years of the journal reveal that in many instances it is applied not just to the cutting of trees from forests, but of the *cutting* of logs into various products, as in a recounting of the activities of the School of Forestry's annual spring camp; 'tough English oak, sixty years old, to be felled and cut into posts; springy, gummy young pines … to be cut up for cordwood' (Anon, 1930).

Logging as a verb has, as its base word, the object of the doing – the *log*. This would be a logical change in terminology for a stewardship forestry approach. The forest is no longer to be *cut*, it is to be retained. Only the log is to be removed, so wood-getting is now restrained to *logging*, rather than the final, and now considered wasteful, *cutting*. *Harvesting* is the most recent of the three terms to have a period of positive specificity. It is agricultural in evocation. Crops are *harvested*. With the emergence of crops of trees the term *crops* itself has its highest specificity in the period of the 1970s and early 1980s – at the start of the rise of the term *harvesting*. The most common co-occurrences for *crop* point to the word being used primarily in relation to trees (for example, *tree, trees, nurse [crop], rotation, thinning* and *radiata)*. The rise in usage of the word *crop* and *harvest* is clearly evocative of the new agricultural wood cultivation basin. The increase in the use of the word *crop* occurs at a time when pine production is growing rapidly in New Zealand, but before the emergence of growing concern for ecological considerations in forest management. As foresters increasingly tackle the public demand for non-wood values in forests, it is possible that a nascent conceptualisation of forest as crop was deemed incongruous with emerging ideas of forests as complex natural systems, a view given additional impetus by a growing call for sustainable forest management. The wood/forest nexus unravels as tensions within forestry, between managing forests as complex natural ecosystems that provide diverse values and managing *crops* of trees as specialised wood cultivation systems, render it unnecessary.

Another reason for the shift to the word *harvest* is as a response to the growth of environmental demands. If the conservation movement was successful in shifting public opinion against *logging* in natural forests, it could have been advantageous to shift the term to *harvest* – a word evocative of a more widely accepted process of growing food in monocultural crops, but also evocative of the *harvesting* of, say, mushrooms or seeds from a forest – a largely benign undertaking. Perhaps *logging* had come to be seen as too harsh an intrusion into the forest while *harvesting* could be conceived as an acceptable part of managing a natural forest – an evocation of light-touch wood removal that largely leaves the forest intact. This is a type of wood removal that retains the whole forest ecosystem as sought through ideals of sustainable forest management and other adaptive approaches.

The shift in terms could also represent a shift to an increasingly technological form of converting trees to wood. To *cut* once is just to start the removing of a single log, while to *log* is to remove an entire tree's log, all of its wood, in one act. It is an escalation of the impact over cutting. *Harvest* goes beyond the getting of one log to the collection of all the logs in the one act: '[a]ltogether this group represents some 90,000 hectares of planted production forests with an annual harvest of some 1.3 million cubic metres' (Anon, 2002, p. 33).

A *cutter* cuts wood with an axe, a *logger* logs trees with chainsaws and heavy machines, while a *harvester* harvests trees with automated harvesters. Indeed a *harvester* is more commonly understood to be a machine, not a person like a *cutter* or *logger*. A wood *cutter* is evocative of an axe-wielding human. A *logger* invokes a hard hatted chainsaw wielder backed up with heavy machinery. A *harvester* however, is a more generally associated with a machine that does the *harvest* and retrieval of the *crop*. If a person is involved it is an operator, someone who controls the harvester. The efficiency and extent of the impact grows with each change but with *harvest* concerns about this are moderated through its other benign agricultural associations. *Harvest* perhaps does a better job than *log* in that it invokes a sense that the trees will grow back easily, naturally, reliably.

Discussion

The text analysis shows word use change that reflect the model of the transition from stewardship forestry to forest ecosystem management and wood cultivation. The use of words related to forest ecosystem values such as *climate*, *biodiversity* and *sustainability* all increase towards the end of the period, reflecting the rise of social demands for these values in forests. In addition, terms such as *conflict* or *parks* peaked during the period of the 1980s and 1990s, coinciding with the period when discussion about the shift of wood production out of remaining natural forests was occurring. Attempts at adaptation within the stewardship forestry basin struggled with the transformative pressure to shift wood production to plantations alone. The evolving trend in the use of the words, *cut*, *log* and *harvest* is indicative of the progression of wood production from exploitation of endowment forests through stewardship forestry to wood cultivation. The following discussion considers the implications of these observations for the model of change in wood and forest social-ecological systems with attention on what this says about the institution of the professional forester.

As might be expected, the journal set up to represent the profession that existed to save forests from exploitation makes clear reference to the legacy of the exploitation phase. It was observed that some wood types were 'liquidated' to such an extent that attempts at sustainable forestry were simply too late (McKelvey, 1987). Equally, the potential to remove natural forest from the foresters remit through protection in reserves was talked about decades before the peak of the changes in this regard occurred in the 1980s. As with the analysis of Indonesia, this is suggestive of a lack of neat transition in social-ecological system states. Rather, these shifts are messy.

As a further example, the discussion about New Zealand's shift to plantation and, hence, the wood cultivation basin in the early decades of the twentieth century, even while establishing a stewardship forestry basin, is telling. While some foresters worked to establish sustainable systems for New Zealand's natural forests, others were working to set up a new wood cultivation regime. The annual reports of the first New Zealand Director of Forestry, MacIntosh Ellis, in the 1920s were said to 'reflect the natural instinct of the forester to rely upon the indigenous stands as his basic source of future timber supplies', a stewardship forestry approach, 'and it was only with much reluctance that he abandoned this idea as an impractical solution to the country's future timber problem,' to become a champion of the major exotic pine plantation schemes of the 1920s in New Zealand (Anon, 1944, p. 8). The passage notes that even as the formal institutions of stewardship forestry were first established in New Zealand there was recognition that the remaining indigenous forests alone were not capable of meeting future wood demands and that wood cultivation was needed. It might be, then, that the transition between basins is not a simple one of sequential transformation or adaptation, but that two basins of the same system can coexist during the change. One basin does not emerge into being after the other passes away; rather, different basins as states of the same system might coexist for a period, with one or the other coming to dominate at some point.

In part, this highlights the panarchy nature of these systems, the arbitrariness with which we define them, as well as the simple state flip conceptualisation. However, the situation described here for New Zealand as a single wood and forest social-ecological system suggests that for the period studied at least it was in a continual state of shifting through states, and never actually settling into a fixed stable basin. Here, then, is an element of arbitrariness not just in defining the social-ecological system but also the particular basin. Finally, by defining the social-ecological system at a national level we end up observing a system that is itself an amalgam of smaller social-ecological systems. It is unlikely that all the social-ecological systems within the larger one will exist in the same basin at the same time. For example, if some districts have lost most of their natural forests but have considerable plantations, then the district as a discrete wood and forest social-ecological system can be said to have transformed from one of stewardship forestry to wood cultivation. Again, we see the panarchy effects operating. Change in one district is likely to impact upon other districts – for example, through a transfer of knowledge of steep slope planting techniques, or improved pine cultivars, or new processing plants for plantation pine shifting demand in favour of plantations in neighbouring districts.

This conceptual intergrading can also be seen in the overlap in skills and interests as forest management shifts from foresters to restoration ecologists and protected area managers. Sarr *et al.* (2004) ask, '[w]here does silviculture end and restoration ecology begin? Are they different approaches on a continuum of management options, or do they represent different spheres entirely?' These questions point to the challenge of defining separate states, but the answers might also depend on the motives of the respondent and where they are sited in the contestation of visions and modes of description of social-ecological systems. Who defines these systems, and why, is important. For example, foresters might be more inclined to answer that restoration ecology is just a recent adaptation of forestry thinking, while restoration ecologists might argue for their distinctiveness:

> ... the primary distinctions between restoration ecology and forestry lie in aim and scope. Restoration ecology typically places ecological goals ahead of economic goals and poses different questions and treatments than most forest management approaches.
>
> (Sarr *et al.*, 2004, p. 21)

The profession of forester is a central institution of the stewardship forestry basin. It is useful to consider what this analysis might say about the potential 'performativity' of this profession. Callon (2006), using the example of economists, speaks of performativity as the way in which economics 'perform' markets; that is, how economists affect the subject of their study. MacKenzie (2006) notes that while it is obvious that economics will be used in economic practice, where it becomes 'interesting' is when what he calls 'effective performativity' takes place. This is when the system itself (markets in MacKenzie's example) are affected by those who perhaps mistakenly study the system as an object somehow unaffected by their study. This does raise the question then: do foresters perform forestry?

Has forestry ever been a pure science with a capacity for objective distance from its subject? Or is it an applied science with performative effects or even explicit normative aspirations? The latter is a reasonable assessment, given the overarching imperative on which stewardship forestry institutions were founded – to produce sustainable wood supplies through effective forest management. In this sense the performativity of forestry as a science is less surprising then the 'outing' of economics by scholars of science (MacKenzie, 2006). Instead, foresters always intended to alter the system they study. They have, then, been engaged in what Latour (1987) calls the task of *enrolling* others in a project to build its reality as fact. This in turn makes their task more clearly one that is both politically and socially constructive. It is clear that foresters, as a profession, have always worn their normative aspirations on their sleeves, as it were.

Arts *et al.* (2014. p. 7) describe this performative function of forestry when they state that, 'in practice, the articulation or the enactment of discourse allows for constant modifications and adjustments. In other words, knowledge and reality constitute each other ... one cannot be taken to pre-exist or cause the other.' Thus knowledge and discourses shape and inform social-ecological systems, including their social institutions, which in turn influence future knowledge production. Consistent with this, the foresters' development at the nexus of wood and forests would have set up a performative role that acts to perpetrate this nexus, despite change in the state of the system no longer requiring it. Foresters would act to maintain the system in the face of pressures that would otherwise have changed it, even when there are compelling new logics. Yet, as we have observed, this social-ecological system is subject to powerful forces from outside. It is not just foresters who create knowledge that effects the reality of wood and forest social-ecological systems.

Consider the centrality of the term *forest* in defining forestry. What happens when forests are parts of landscapes where forest intergrades with non-forest, such as when in conservation areas forests intergrade with say, grasslands, savannah and wetlands. Where these forests are managed as part of such whole landscapes for non-wood values they are often not managed by *foresters* or *forestry*. What is there to distinguish the forester from the conservation reserve manager, the natural resource manager, the plantation manager, the agronomist or the farmer? The evidence points to wood production. They might then have been more appropriately named *wooders*, but of course they took the name when the wood/forest nexus made *forestry* perfectly logical. Kirkland (1984) observed that '[e]ventually the distinction between forestry, agroforestry and pastoral farming must become progressively eroded', and this analysis supports such a position. The weight of evidence suggests that the *forester* and *forestry* (the terms are strongly tied) are centred on the wood/forest nexus. As this wood/forest nexus comes undone (as wood production shifts from extensive forest management to intensive cultivation of trees for wood) we see the emergence of a post-forestry state in wood production and forest ecosystem management.

Notes

1 From 1925 until 1936 the journal was published as *TeKuraNgahere*, which started in 1925 in the Canterbury College of Forestry. It became the official journal of the New Zealand Institute of Foresters from 1934 and from 1937 the journal was named *The New Zealand Journal of Forestry* (between 1986 and 1998 it went by the shorter name of *New Zealand Forestry*).

2 Random sampling found about 1 per cent of words in the text files were incorrect interpretations of the original files by the optical character recognition software used. It was not considered that these errors would alter the broadly observed trend patterns used in the analysis here.

3 A 'word' here is a particular set of case specific letters. So, for example, *Forestry, forestry* and *FORESTRY* are each treated as separate 'words' in this count.

References

Anon (1930). Camp notes. *TeKuraNgahere*, 2, 46–47.

Anon (1935). Abstract of the annual report of the director of forestry, New Zealand State Forest Service. *TeKuraNgahere*, 3, 234–236.

Anon (1944). Obituary: L. MacIntosh Ellis. *The New Zealand Journal of Forestry*, 5, 6–8.

Anon (1946). Editorial – Waipoua Kauri Forest. *The New Zealand Journal of Forestry*, 5, 173–175.

Anon (1988). Institute's new President – Dr W.R.J. Sutton. *New Zealand Forestry*, 33, 26.

Anon (2002). New group raising Otago/Southland's forest products profile. *New Zealand Journal of Forestry*, 47, 33.

Arts, B., Behagel, J., Turnhout, E., De Koning, J. and Van Bommel, S. (2014). A practice based approach to forest governance. *Forest Policy and Economics*, 49, 4–11.

Baddeley, C. (1993). Recent events: Forest Amendment Act passed. *New Zealand Forestry*, 38, 10.

Bell, M. (1995). Our forests forever?: language and ideology in community education materials. *Australian Journal of Adult and Community Education*, 35, 219.

Burdon, R.D. (2008). Breeding radiata pine – future technical challenges. *New Zealand Journal of Forestry*, 53, 3–5.

Callon, M. (2006). What does it mean to say that economics is performative? *CSI Working Paper Series*. Centre de Sociologie de l'Innovation.

Charteris, P., Jack, M., Cown, D., Clinton, P. and Barnard, T. (2010). New Zealand speakers. *New Zealand Journal of Forestry*, 55, 3–5.

Clark, A.F. (1926). Canterbury's native bush. *Te Kura Ngahere*, 2, 15–22.

Drouin, P. (2004). Detection of domain specific terminology using corpora comparison. *Proceedings of LREC*. Lisbon.

Dunlap, R.E. and York, R. (2008). The globalization of environmental concern and the limits of the post-materialist values explanation: evidence from four multinational surveys. *The Sociological Quarterly*, 49, 529–563.

FAO (2014). Country report: New Zealand. In: Forestry Department Food and Agriculture Organisation of the United Nations (ed.) *Global Forest Resources Assessment 2015*. Rome: Food and Agriculture Organisation of the United Nations.

Fleet, H. (1984). *New Zealand's forests*. Auckland: Heinemann.

Giddens, A. (1990). *The consequences of modernity*. Cambridge: Polity Press.

Goulding, C. (2013). New Zealand is the world's second largest exporter of unprocessed logs – this is not a compliment. *New Zealand Journal of Forestry*, 58, 44.

Halkett, J.C. (1987). Transparency of purpose: an improvement on the Forest Service? *New Zealand Forestry*, 32, 20–23.

Halkett, J.C. (1989). Foresters to forestry – one year on. *New Zealand Forestry*, 34, 29.

Hillier, J. (2003). Fighting over the forests: environmental conflict and decision-making capacity in forest planning processes. *Australian Geographical Studies*, 41, 251–269.

Holloway, J.S. (1983). Editorial. *New Zealand Journal of Forestry*, 28, 1–6.

Jackson, D.S. (1964). Editorial notes. *The New Zealand Journal of Forestry*, 9, 5–6.

Kirkland, A. (1984). Guest editorial. *New Zealand Journal of Forestry*, 29, 1–5.

Latour, B. (1987). *Science in action: How to follow scientists and engineers through society*. Cambridge, MA: Harvard University Press.

Leslie, A. (1976). The environmental challenge to forestry. *New Zealand Journal of Forestry*, 21, 5–16.

Loughlin, C.O. (1990). Forest research – coping with change. *New Zealand Forestry*, 35, 16–19.

Mackenzie, D. (2006). Is economics performative? Option theory and the construction of derivatives markets. *Journal of the History of Economic Thought*, 28, 29–55.

McKelvey, P.J. (1987). World of the Kauri. *New Zealand Forestry*, 32, 37.

Mead, D.J. (1986). Editorial. *New Zealand Forestry*, 31, 2.

Mebratu, D. (1998). Sustainability and sustainable development: historical and conceptual review. *Environmental Impact Assessment Review*, 18, 493–520.

Moore, P. (2001). Biotechnology: global aspects of the debate on biotechnology. *New Zealand Journal of Forestry*, 46, 10–12.

Poole, A.L. (1959). Editorial Notes. *The New Zealand Journal of Forestry*, 8, 1–8.

Roche, M.M. (1987). *Forest policy in New Zealand: an historical geography, 1840–1919*. Dunmore Press.

Rockwell, J. (1982). President's address 1982. *New Zealand Forestry*, 27, 23–26.

Sarr, D., Puettmann, K., Pabst, R., Cornett, M. and Arguello, L. (2004). Restoration ecology: new perspectives and opportunities for forestry. *Journal of Forestry*, 102, 20–24.

Treasury. (2015). *New Zealand Forest Research Institute Ltd* [Online]. Wellington. Available: www.treasury.govt.nz/statesector/commercial/portfolio/bytype/cris/new-zealand-forest-research-institute [Accessed 3 August 2016].

Van Dijk, T.A. (2001). Multidisciplinary CDA: A plea for diversity. In: Wodak, R. and Meyer, M. (eds.) *Methods of critical discourse analysis*. 2011 ed. London: Sage Publications.

Westoby, J.C. (1970). One-World Forestry: New Zealand's role. *The New Zealand Journal of Forestry*, 15, 9–24.

Wetherell, M. (2001). Themes in discourse research: the case of Diana. In: Wetherell, M., Taylor, S. and Yates, S. (eds.). *Discourse theory and practice: A reader*. London: Sage Publications.

Wijewardana, D. (2005). Market perspectives for timber from planted forests. *New Zealand Journal of Forestry*, 50, 34–41.

Wolf, S.A. and Klein, J.A. (2007). Enter the working forest: discourse analysis in the Northern Forest. *Geoforum*, 38, 985–998.

Xu, Z. and Bengston, D.N. (1997). Trends in national forest values among forestry professionals, environmentalists, and the news media, 1982–1993. *Society & Natural Resources*, 10, 43–59.

Part III

Transitions and resilience

8 Wood and forests in a post-forestry world

Global wood production from natural forests peaked around 1989 and has been in decline since (Figure 8.1). As was noted, this global pattern reflects that shown in many individual countries – it is being replicated at different scales and in multiple locations. These results describe a non-sustainable resource use pattern for wood extraction from the world's natural forest ecosystems. As noted earlier, global wood production can only said to be sustainable by incorporating wood production from cultivated wood sources. With this shift to cultivated sources wood production is concentrating into a smaller area of production. A conservative 33 per cent of the world's industrial roundwood production was produced from the 3 per cent of the world's

Figure 8.1 Total global roundwood production (industrial and fuel wood) 1945–2016 and cultivated roundwood and natural forest roundwood portions of total roundwood.[1]

'forest' that were productive plantations in 2012. On a per unit area basis these productive plantations were 18.7 times as productive as the world's natural forests and 2.6 times as productive as the world's semi-natural forests (see Figure 2.2).

But these findings in turn raise the question: So what? What does this mean for wood and forest social-ecological systems? And what does it mean for other resource use systems and how they respond to change? Answering these questions has significance for our understanding of processes of change (theory) and for recommended responses to change (policy). This chapter will focus on the implications of these changes for wood and forest social-ecological systems. It starts with a review of key theories of land sparing/sharing and the potential for a *plantation conservation benefit* (the potential for an expansion of wood cultivation to take wood harvesting pressure off natural forests). This is followed by consideration of the future of stewardship forestry in response to changes in its conditions.

First, however, it is worth recapping the process of change outlined in wood and forest social-ecological systems. Four basins of the wood and forest social-ecological system were defined. Each had an underlying logic and there have been internal contradictions to this logic, pressures that have given rise to eventual transformation to a new basin. In summary, for forest exploitation the logic is simply the need for wood and meeting this need through the exploitation of trees in extant natural forest. Its chief challenge is that the rate of extraction in a growing and or industrialising population can easily outstrip the replenishment of wood in those forests – its unsustainability. For stewardship forestry the logic of the basin is to achieve this sustainability, to steward the forests so that wood is only supplied at a level that can be sustained. The challenge for stewardship forestry comes from economic pressures acting to drive wood production out of the forests to cultivated wood sources (and also reducing demand through increasing efficiencies in wood use), and social and ecological pressures to optimise non-wood values from natural forests challenging the value of wood production in natural forests.

In response to these challenges, wood production and forest ecosystem management are separating into two new basins. For wood production this is the wood cultivation basin and its logic of optimal, and sustained, wood production – the logic of agriculture. It combines both the stewardship approaches of sustainability and modernity's drivers of constant technological and social change to underpin efficiency and growth. For forest ecosystems this is the forest ecosystem management basin. It shares the sustainability logic of stewardship forestry, the maintenance of forests, but without the basic rationale of wood production. These later two basins are likely to be confronted with crises of their own in the future although these are not assessed in this book – the focus is on the stewardship forestry basin. The implications of this shift are the subject of larger discussions about optimal land use strategies for forests, wood production and a range of other land uses across the landscape.

Land sharing, land sparing and the *Plantation Conservation Benefit*

In forest management discourse there are tensions within and between the older regimes of stewardship forestry and emerging wood cultivation and extensive forest ecosystem management approaches about optimal forest management approaches (and indeed broader landscape management approaches). A common line of questioning goes like this: Should there be less wood harvest in natural forests in order to reduce impacts of wood production on other forest values? Should wood production be focused in small areas of highly productive plantation to take pressure of natural forests? Or, is it optimal to integrate wood production into natural forest ecosystems to help effectively sustain and manage these? These questions overlap with the debate about land sparing and land sharing in conservation management (Fischer *et al.*, 2011; Phalan *et al.*, 2011). *Land sparing* is where a natural ecosystem is spared the impact of a particular use by meeting the demand from a smaller area of more intensive production/extraction/use. The counterargument is that management of natural ecosystems provides for multiple benefits arising from the integration of different land uses (including extractive ones) such as a higher capacity to absorb impacts, greater ecological resilience than monoculture systems or income from suitable light footprint extraction that helps pay for the management of other conservation values that are harder to monetise. This constitutes *land sharing*.

Overall, the global transition of wood production from natural forest extraction to wood cultivation and a shift of natural forest management away from wood extraction is land sparing writ large. This raises interesting questions for the debate as to which is better – sparing or sharing. Often these are spoken about in normative terms, however some of the rationale for the transition in wood production outlined here suggest that a land sparing outcome arises out of processes occurring outside the remit of foresters, conservationists or forest ecosystem managers – processes of technological change and economic efficiency and profit drivers for example. This, then, raises the possibility for a positivist analysis. To what extent is a land sharing or land sparing outcome a logical outcome of circumstances? Certainly, evolutions in wood and forest social-ecological systems are tending towards a land sparing condition, even if that is not always a deliberate normative aspiration.

However, thinking of the transition from stewardship forestry to wood cultivation and forest ecosystem management as land sparing alone is too simple. It also displays land sharing qualities in a number of ways. First, not all wood cultivation occurs as industrial monoculture. As outlined in the chapter *Global analysis of trends in wood sourcing*, there has been a significant increase in wood production from trees not from forests, in particular in the tropics in response to demand for wood fuel. This is a significant portion of the world's total wood production – in the order of 18–35 per cent – in estimates in that chapter. Here, the trees are often grown as part of agroforestry landscapes, as smallholder woodlots or as trees planted interstitially in existing rural landscapes. These then constitute a form of land sharing where the wood production is integrated into rural and agricultural landscapes. In addition, in landscapes that see pasture or other monoculture crops replaced with plantation, monoculture tree plantations can be managed in ways that lead to improved biodiversity outcomes, especially when mosaic landscape approaches are used (Brockerhoff *et al.*, 2008). Second, while removing wood production from natural forest estates to achieve greater conservation outcomes can be labelled as a simplification of land use (the 'locking up' of forest [for example Bengston *et al.*, 2001; Rasker and Hackman, 1996]), contemporary active management of protected areas for diverse ecosystem services can involve a greater diversity of land use than earlier stewardship forestry regimes that were focused on wood production alone (an 'opening up' rather than 'locking up' of forest). The shift of wood production can be seen as a migration of the singular function of wood production out of natural forest landscapes into rural landscapes driven by compelling logics. While this alters these landscapes it does not necessarily compromise the complexity or multi-functionality of either.

There can be no doubt that at a global level the world is shifting to a more land sparing approach in relation to wood production and forests. The pattern of increasing land sparing is reflected in the three case study countries. This idea of a land sparing approach in forestry has been around for some time. As noted earlier, numerous stewardship foresters have recognised the quandary presented by limitations on the capacity of remaining natural forests to meet timber demand, while satisfying increasing demands for forests to be set aside from wood harvest in order protect non-wood values. Further, land sharing can struggle to make its case at a global level where the mismatch between global demand and available forests becomes clear. Fox (2000) observed that certified natural forest wood production is in the order of $0.7\,m^3/ha/yr^2$, which, he noted, would have taken 4.7 billion hectares of natural forest to supply all the world's wood demand at the time. In 2015 the world had 3.7 billion hectares of natural forest – and 1.3 billion of that was primary forest (FAO, 2015). A significant gap exists between global demand and the ability of the world's remaining forest ecosystems to meet that demand when applying such sustainable stewardship forestry approaches. This highlights the difficulty of delivering wood demand from land sharing approaches (such as sustainable forest management, ecological forestry, restoration forestry and third party certified forestry). Although, this is not to say that there is no capacity for a mix of both approaches as occurs now.

As a result many have embraced the solution of using high productivity wood plantations. The idea has a 'logical simplicity' that makes it 'not surprising that this idea has a long lineage' (Pirard *et al.*, 2016, p. 122). An extension of this idea is that setting up plantations will cause wood production in natural forests to be reduced – the *plantation conservation benefit*. It is the same logic the Borlaug hypothesis proposes in relation to agriculture – intensification of agriculture can save forests by increasing agricultural production from the same, or an even smaller area of land, thus negating the pressure to increase agricultural land at the expense of forests (Angelsen and Kaimowitz, 2001). Applied to forests and wood production, intensive wood cultivation takes the place of agricultural intensification and the subsequent prevention of forest degradation through logging.

When considering the plantation conservation benefit there is an obvious correlation between increased cultivated wood production and growth in the forest conservation estate. But making causative links is problematic and it is difficult to determine the extent to which wood harvest from plantations might actively increase natural forest protection (Pirard *et al.*, 2016). It is difficult to determine, for example, the effects of subsidies on the price of both plantation and natural forest-sourced woods, or the extent to which plantations are developed in response to declining natural forest wood supplies. Nor is it clear that policy decisions have been the driving cause of the plantation conservation benefit. In the Australian case, for example, it is unclear that it was a policy intention to deliver a plantation conservation benefit. Rather, other policy intentions have been posited, such as encouraging plantations to grow overall wood production (with no intent to increase natural forest conservation), or encouraging plantations expansion in response to anticipated mismatches between future demand and natural forest supply.

As well as these problems of showing causation, the plantation conservation benefit hypothesis often relies on a number of assumptions, as noted by Pirard *et al.* (2016). One assumption is that logging forests causes a loss of conservation values. There is evidence for this, despite substantial effort to achieve sustainable forest management, as noted earlier, but it is far from conclusive. A second assumption is that all natural forests have equal conservation benefit. It is difficult enough to determine the economic effects of a particular set of plantations on natural forests as a whole, let alone comparing the values of different forests. The use of this assumption was a limitation noted in the chapter analysing leakage in Australia. Assumption three is that wood from plantations can substitute wood from natural forests. Wood types from plantations and natural forests are rarely perfect substitutes. However, changes in wood processing technologies are often able to adapt to these imperfect substitutes. This is becoming more so as wood processing shifts to engineered and reconstituted wood products where differences in timber qualities can be less significant. A fourth assumption is that the effects of leakage are known and managed. Trade leakage can undo the conservation benefits of forest protection in one country. The Australian case showed that the plantation conservation benefit can occur without leakage, but there are contradictory examples from other countries such as Vietnam (Ingalls *et al.*, 2018). This highlights how these ideas can suffer from the difficulty of translating a process that is clear at one scale but not when observed at others.

Concern about leakage in response to proposed actions to protect and conserve forest is valid, although observations such as '[l]ogging bans for instance usually transfer harvest pressure onto neighbouring areas … and so achieve nothing at the global level' (Fenning and Gershenzon, 2002, p. 295) is extreme and doesn't match with global level evidence. As noted in the Australian case there is considerable scope for forest protection to occur in alignment with new plantation establishment. The observation above also ignores the induced innovation in wood use efficiency that can offset reduced wood supply and, indeed, the role of restricting wood supply through forest protection in inducing this efficiency. When considering increasing conservation of forest ecosystems through reduced logging, policy makers need to be cognisant that this policy option can act to induce innovation in both wood cultivation and wood processing (see also the Porter hypothesis in the next chapter). Thus forest conservation policy can have wood industry outcomes other than leakage of wood production to other forests.

Further complicating the policy calculation, Pirard *et al.* (2016) note two possible negative outcomes for natural forests associated with increased plantation wood. First, there is the possibility of rebound effects (Greening *et al.*, 2000) along the lines of the Jevons paradox (Jevons, 1865) occurring in relation to wood demand. This says that as the price of wood falls due to increased plantation wood supply, demand grows and pressure *increases* on natural forests. A consideration here is the extent to which plantations act to lower wood price, thus setting up conditions for rebound effects. If plantation wood supply is being developed because higher wood prices make it increasingly viable, this is less likely to produce a rebound effect. In this context adequate regulation and enforcement of sustainable yield also become critical and can provide protection against the possibility of this effect. Also, it needs to be noted that the Jevons paradox is based on a perfect substitute of commodity coal. Wood from natural forests and wood from plantations are not so perfect substitutes as coal. This is likely to also limit the potential rebound effects.

Another significant challenge to the plantation conservation benefit raised by Pirard *et al.* (2016) is the possibility that lowering the value of natural forest land by reducing demand for its wood renders it more prone to being cleared for other land uses. While the corpus studied in the literature review by Pirard *et al.* (2016) did not contain specific cases of this, the theoretical modelling points to it being a possibility. But this concern rests on land (and its forests) being privately owned and capable of being sold in something like a perfect market. However, around 80 per cent of the world's forests are owned by governments and are not generally available for trading in markets. They are subject to government regulation (when well enforced) although, conversely, problems of commons (where regulations are poorly enforced) can act in manners that reflect those market concerns noted above. In both cases – rebound effects and deforestation effects – a key need is for a combination of policy to support plantation expansion and policy to protect natural forest from wood harvest. In the three national case studies presented here, government policy has played a significant role in determining how wood and forest social-ecological systems have been shaped – something that is missed when analysing land use patterns through market analyses alone.

The policy mechanisms displayed in the case studies include protection of forests in conservation reserves, regulatory restrictions on harvesting of timber and reviews, and reductions in allowable annual cuts. Importantly, these restrictions can have the effect of reducing wood supply and therefore triggering the same sort of economic processes that a complete exhaustion of endowment or natural forest wood supply will have – induced innovation and investment in wood cultivation increasing its comparative advantage. A clear synergistic advantage exists to having in place a combined wood industrial policy and forest management policy to optimise the potential of achieving a plantation conservation benefit.

As noted in the Australian case study, this sort of integrated policy is also critical to reduce or lessen the potential impact of leakage effects. As noted in the case of Australia, there has been a synchronous arrangement of plantation policy increasing the plantation estate and its productivity while increasing the area of natural forests protected from wood harvest and managed for non-wood values. This experience is relevant to other nations facing combined pressures to conserve natural forest and develop plantations and other wood supply alternatives. The New Zealand case outlines a similar example. Further, in the case of Australia and New Zealand the net effect of the transition to a cultivated wood products sector could be net positive leakage as they have become significant exporters of cultivated wood into global markets.

While the rationale for the shift to wood cultivation is compelling, it needs to be recognised that there are real and significant challenges in further enlarging wood cultivation if the process is to continue. The increasing role of cultivated wood in roundwood production includes real threats to biodiversity conservation, especially where conversion of natural forest is involved. To address these threats, advice to cease conversion (for example, Elias *et al.*, 2012) must be heeded, as well as making use of the opportunities for existing and new cultivated wood plantings to deliver a range of ecosystem services, especially where they are established on previously degraded, abandoned and barren landscapes (Bauhus *et al.*, 2010; Silva *et al.*, 2018). In addition to these ecological impacts there are significant social impacts resulting from the rapid landscape changes that wood cultivation entails, especially when coming from industrial plantation. Recent analysis from Chile (Andersson *et al.*, 2016), Indonesia (Pirard *et al.*, 2017) and China (D'Amato *et al.*, 2017) highlight a mix of opportunities and problems associated with large scale development of industrial plantations. Notably, all three sets of research recommend that companies undertaking industrial plantations have the capacity to mitigate these effects through appropriate engagement with local communities in coordination with government. Achieving a plantation conservation benefit needs to not only take pressure off natural forests but in establishing plantations to seek to deliver conservation (as well as social) benefits in broader landscapes.

The above review of the *plantation conservation benefit* raises another interesting possibility – the potential for social values to be realised to a greater extent than would be possible in a wood production regime. In particular, there has been much written about the need to consider social forestry in managing forests, including in the chapter on Indonesia in this book. Like forest conservation values, achieving social forestry values in natural forests can be challenged by wood production interests. By shifting wood production to wood cultivation sources, might there be better realisation of social forestry outcomes? This has particular potential in places like Indonesia where there are often intense conflicts between social groups with strong traditional connections, with forests having access restricted or denied by state and corporate interests as the natural forest is utilised for wood production. This, then, suggests the possibility of a *plantation social benefit* accruing from new wood cultivation in similar ways to the plantation conservation benefit.

The future of stewardship forestry

Fundamentally, stewardship forestry has been about the pursuit of sustainability of wood supply. So, has stewardship forestry succeeded in its goal of sustainability? Or has it largely been a form of controlled exploitation of endowment forests? Certainly the institutions of stewardship forestry are correctly credited with having established the principles of sustainable natural resource management well in advance of its more general uptake in the late twentieth century (Wiersum, 1995). However, in the specific case of forest ecosystems much of the wood production in the world's endowment natural forests has been due to the exploitation of primary forests or successive cuts for lower value logs. Inasmuch as stewardship forestry has overseen this, it reflects a managed extraction of endowment forests rather than a perpetually sustainable forest management. As was noted earlier, the earliest wood extraction often occurred in parallel with land clearing. This was followed by a period in which wood exploitation moved beyond the land clearing frontier and took on the character of selective logging for high value species, and high grading[3] (Hyde, 2012, p. 179). Subsequently, forests were logged with new silvicultural techniques, for example using clearfell and fire to increase regeneration of preferred wood species, often with the consequence of ecological simplification (Lunney, 1987). Each stage can involve the timber industry returning and adopting to smaller and lower grades of log (Hyde, 2012, p. 211).

Of the case study countries, Australia comes closest to displaying a sustainability from stewardship forestry applied to natural forests, with a consistent supply of natural forest wood over several decades (at least until the start of the Regional Forest Agreement period that commenced in the later 1990s; see Figure 1.3). This suggests a form of extensive natural forest management that maintained wood productivity on a sustainable basis. Yet, even here forests continued to degrade, particularly through diminution of forest diversity and ecological maturity (see, for example, Bekessy *et al.*, 2009; Lindenmayer *et al.*, 2012; Lunney, 1987). While stewardship forestry approaches have been significant in the development of institutional arrangements for forest management and wood production, it is less likely that sustainable stewardship forestry over multiple rotations in natural forests has had a significant effect on the production of the world's wood, possibly with the exception of localised examples and some European forests. Rather, much global wood extraction has been an opportunistic activity based on accessing forests with large standing volumes either through untrammelled exploitation or some form of moderated exploitation overseen by institutions of stewardship forestry. Large volumes of endowment wood have been taken from forests with little or no history of stewardship forestry silvicultural management in the past. If the shift to wood cultivation continues as it is then little of the world's wood will ever come from natural forests managed by long histories of stewardship management.

The experiences of Indonesia, New Zealand and Australia all support this idea that stewardship forestry has acted as a form of control or brake on the exploitation of endowment forests, but has generally been unable to deliver the goal of a perpetual sustainability. That is, the forests have continued to be utilised in a forest exploitation manner, albeit one that is controlled and slower than might have otherwise occurred. Part of the reason for this is also that the stewardship forestry basin has had to contend with the pressures of rising demand for ecological services other than wood from the managed forests – as well as the burgeoning global populations, shifting wood demand and changing technologies that triggered the need for stewardship forestry in the first place. The story of stewardship forestry then is very much one of reactive response to dramatic change. The goal of the establishment and maintenance of an enduring regime of sustainable equilibrium in forest and wood production systems has not been possible in these conditions.

This gap between the ephemeral systemic change stewardship forestry has faced and its aspirations of sustained stability can be seen in how it grapples with a distinctive feature of wood production – the particularly long production cycle for its wood (Binkley, 2003; Hyde, 2012). This is more pronounced when dealing with the generally slower growing trees of natural forest ecosystems. This provides a range of challenges for associated industries, such as risk management challenges, difficulties in attracting investment and the challenges of foreseeing future market conditions. The forest industry, then, has strong incentive to take advantage of standing stocks that exist now rather than engage in the risks of investing in future wood sources. This could partly explain why there has been such a widespread failure to implement successful stewardship forestry in natural forests. It also explains the attraction of much shorter rotation wood fibre cultivation in industrial plantations.

That said, it is feasible that in some forest types, such as the eucalypt forests of Australia or the pine forests of the northern hemisphere, forest management with a sustainable output of 'fibre' is possible. However, as noted above this will likely come at the cost of ecological impacts such as a change in forest age class and structure over time or costly restrictions on harvest so that its economic viability is threatened. These changes run contrary to the expectations of an ecologically conscious citizenry. This demand for maintenance and even enhancement of ecological diversity significantly limits the capacity to optimally manage natural forest for wood or fibre production. Even in Europe, with relatively long histories of stewardship forestry, there is a shift to a greater differentiation between forest ecosystem management and wood production forests in order to achieve greater ecological outcomes (BMU, 2007; Meyerhoff *et al.*, 2012).

Changes in wood production and forest management social-ecological systems is a notable feature of their long history. Yet, paradoxically, the historically extensive realms of forests, the long timeframes of forest and slow wood growth, and the apparent stability in patterns of use across generations has contributed to strong notions of stability and equilibrium in the institutions of forestry. As Kant (2000) notes, this fundamental gap between the practice and culture of forest managers with approaches of stable biology and equilibrium economics are fundamentally at odds with the highly directional and evolutionary nature of the socio-technological-biological evolution of forests, wood and the social systems that interact with them.

This tension between stability and change is mirrored in the tensions between the emerging possibility of transformation and the drive for resilience and efforts at adaptation in stewardship forestry. Conservatism arising from the long term nature of stewardship forestry (Glück, 1987), institutional 'stickiness' or inertia from the large public organisations, related distributional coalitions (Munck af Rosenschöld *et al.*, 2014; Pierson, 2000), and adherence to established belief systems all act as attractors contributing to a certain resilience for the stewardship forestry basin. The tension between the resilience generated by these institutional attractors and system stressors is evident, for example in Rist and Moen's (2013) account of the dilemma created by conflict between optimising efficiency in wood production and building resilience in forest ecosystems. Again, they limit the scope of a potential solution by restricting their search for it within the realm of what can be achieved within forests (and forestry). Restricting solutions to adaptation within the domains of forests limits their responses – the potential for transformative change is greatly limited. The possibility of wood production being shifted out of forests and into agricultural landscapes is missed. This blindness to such options in stewardship forestry institutions has been cited as a brake on the uptake of bamboo, a potential substitute for wood fibre in many industrial uses (Buckingham *et al.*, 2014). Likewise, in regard to trees in forests, the 'problematic character of this ... slow growing natural resource' (Lane and McDonald, 2002, pp. 193–194) can be dealt with by fast growing wood plantations (and other fibre sources such as bamboo) – but only if the solutions are looked for beyond the confines of institutional myopia and resistance to transformation.

Of course such resistance is not universal, for example Leslie (2005) hints at, and Paquette and Messier (2009) and Sedjo and Botkin (1997) clearly point to, the emergence of the transformational solution of shifting wood production to cultivated wood and fibre sources outside extensive forest ecosystems. The multiple-use concept in forestry noted above (Rist and Moen, 2013) and changes in forester education (Ferguson, 2012) also demonstrate awareness of changes in the state of wood and forest social-ecological systems and attempts to address this. This adaptability is seen in retention forestry and other ecological forestry techniques. These are designed to allow wood extraction while acknowledging a shift in focus to regenerating forests for their biodiversity and ecological values, rather than simply wood production efficiency.

Another rationale for pursuing such forms of retention forestry is to maintain the income from wood extraction, both to pay for extensive forest management and also to give value to the retention of forest when other land uses are an option. From the perspective of transformative shifts to forest ecosystem management, an alternative way of addressing this is the use of ecosystem service payments. There has been considerable research and trialling of these as a way of deriving income from forests for non-wood values, particularly as public goods (Millennium Ecosystem Assessment, 2005), however broad scale uptake has been limited (Daily *et al.*, 2009; Milder *et al.*, 2010).

In this context, wood extraction might continue to attract interest as a viable source of income to help pay for management, although where strict environmental regulations are in place wood production is still likely to struggle to pay the costs of managing a forest (Oliver and Mesznik, 2006). Zhang (2001) notes that pressure from the high transaction costs of such forms of sustainable wood production in forests predispose the timber industry to large corporate ownership. However, Kant (2000) cautions against any fixed or optimal regime for forest management as this varies with the changing socio-economic dimensions, and 'the conventional view of the superiority of private or state regimes over community regimes has been challenged by a rich body of empirical evidence' (Kant, 2000, p. 287).

Roessiger *et al.* (2011) found that in Germany 'near natural' harvest regimes of selective logging and species diversity could provide an economically optimal choice for risk adverse, small forest owners. Similar approaches in countries with traditional ownership of forests can also lend themselves to forms of community or social forestry (Moeliono *et al.*, 2017; Porter-Bolland *et al.* 2012) where small scale wood harvest can form part of integrated forest management approaches designed to address land rights, rural poverty and ongoing access for non-wood uses of the forests. Yet, with the large scale national and global systems, the balance of social, political and economic factors would seem to favour pressures acting towards wood production from cultivated tree development over small-scale local forest ecosystem stewardship. Such localised stewardship forestry approaches that might work, as noted above, are likely to only ever produce small parts of the world's wood supply and to be driven by considerations other than wood supply.

Another important social consideration from wood and forest management is employment. Stewardship forestry institutions have been credited with steering rural communities in forested landscapes through industrialisation in a relatively good shape (Kennedy *et al.*, 2001). But, overall, technology and resource use efficiency have gone hand in hand with a decline in total jobs in forestry globally. Lebedys and Li (2014) reported that employment in the global forestry subsector (not including the wood and paper processing sectors) had declined from 4.4 million to 3.5 million from 1990 to 2011. Wood processing sectors had also declined, though not as steeply. Understanding the role of innovation in job shedding in forestry is important. The overall role of forestry in the economy of many individual countries has declined, just as it has globally. In Indonesia the forest sector contributed 1.17 per cent to GDP in 2000 and 0.63 per cent in 2013, despite growth in the total wood products sector in real terms over this period. This is reflected in and consistent with global trends over the same period. Jürgensen *et al.* (2014) showed a decline in global GDP contribution by forestry from over 1.2 per cent in 2000 to 0.9 per cent in 2011. This was replicated across all the regions of the world. It continued a similar pattern identified in the previous ten years (1990 to 2000). Particularly when combined with stagnant global demand for wood, these trends should alert policy makers seeking to utilise forestry as a source of employment (especially in rural communities) to apply considerable caution. If worker productivity in wood production continues to increase in line with well-entrenched trends in primary industry and manufacturing sectors, then global wood production is unlikely to deliver overall employment growth (notwithstanding the potential for localised growth as other areas decline).

Future wood demand is another important policy consideration in relation to wood production. A key part of this is making predictions about future demand and supply of wood. Reviewing past efforts reveal how challenging this sort of work is, and how problematic. As noted in the chapter *Global analysis of trends in wood sourcing*, there was a pattern of predicting ever growing demand for wood in the late 1990s and early 2000s that has proved to mostly overstate future wood demand. One reason for this, raised already, is that predictive methodologies have likely failed to take account of changes in technology that deliver more efficient wood use. That is, these models assumed that population and wealth would grow, thus increasing future demand for wood, whilst changes in wood production technologies and wood saving technologies were not adequately considered.

This history of forecast failures also raises the question: to what extent do industry bodies, and perhaps government forecasters including the FAO, tend to 'talk up' future potential growth? To answer this it needs to be recognised that data collection is not purely objective. It is embedded within organisations and their logics (paradigms). This embeddedness acts in two ways. First, there are many questions that can be asked of wood and forest social-ecological systems. But, with limited resources choices are made as to which questions are asked and these are influenced by the values, imperatives and aspirations of those who control the resources to make those choices. An example of this is the question, what portion of the world's wood supply is coming from cultivated sources and natural forest sources? Why was this question not asked and explored in greater depth earlier? Why was there so little data on a major change in wood production? One possibility is that the organisations with the resources to ask these questions were dominated by stewardship forestry thinking and that the distinction (wood cultivation and natural forest) was not of central importance to its main goal of wood production and assumptions of a wood/forest nexus. It is only in recent years that this distinction has started to become a greater subject of research. Of course, it is also the case that the distinction between natural forests and forms of cultivated wood is not always clear. In recent years there have been emerging attempts to address this question (Batra and Pirard, 2015; Carle and Holmgren, 2003; Penna, 2010) and to better clarify categories within global plantation production (Jürgensen *et al.*, 2014). The trend to ask these questions is indicative of the institutions of stewardship forestry beginning to grapple with the shift from stewardship forestry to wood cultivation and forest ecosystem management systems.

But also critical here is understanding that 'official statistics are not just analytical characteristics of social activity, but ... enter constitutively into the social universe from which they are taken and counted.' (Giddens, 1990, p. 42) Such 'authoritative' data is central to state power; indeed, to the constitution of the state. This is also true for the wood and forest social-ecological systems in that the data collected will reflect hegemonic organisations and social networks and act to further consolidate them through a constitutive function. In particular, such data is often accompanied with reports that make normative assertions. Declining wood production data could be either positive or negative, but when collected and reported within wood/forest resource use organisations, whether private or governmental, it is of course likely to be seen as a threat, and communicated as such, influencing and reinforcing the resource use system in one way. Along these lines, there is also 'institutional "gaming"', whereby resource stocks and supplies are overstated for reasons of institutional or organisational advantage (Lindenmayer, 2016). This is possibly behind the preponderance of overestimates in future wood production estimates noted above. Researchers and policy makers, in their work in wood/forest (indeed all resource use) social-ecological systems, need to be aware of the potential for such institutional biases and influences in data collection.

As well as these institutional feedbacks, there are a number of other important considerations in relation to future wood supply predictions. Three noted here are future biofuel demand, increasing CO_2 fertilisation and land and water availability. A major factor in predicting future wood demand will be the role that biofuels play in global energy demand. While traditional levels of fuelwood seem to have peaked and are showing signs of decline, new biofuel demand in a fossil fuel constrained world has led to predictions of rapidly growing future fuelwood production (Buongiorno *et al.*, 2012). Of course, such increases in demand need to be considered in relation to the many processes acting to reduce demand and restrict supply that have already been noted. Use of wood as a fossil fuel replacement in biofuels is also the subject of intense policy and political contention (O'Brien and Bringezu, 2017) further complicating its future prediction.

Increasing carbon dioxide levels will impact on plant productivity over the next century (for example, Kirilenko and Sedjo, 2007; Leakey *et al.* 2009; Taub, 2010). A meta-analysis of 15 papers reporting on long term Free-Air CO_2 Enrichment – in-field fumigation of crops and natural vegetation with carbon dioxide – with summarised findings showing that increased carbon dioxide will stimulate net primary production over the long term along with increased nitrogen efficiency and reduced water demand (Leakey *et al.*, 2009). The evidence in relation to trees and, in particular, forests is ambiguous, particularly due to experiment limitations in controlling all factors (Kirilenko and Sedjo, 2007). However, despite experimental uncertainty about projections, historical work demonstrates that global forest productivity has increased during the second half of the twentieth century. This has been mainly attributed to the warmer climate and a consequent increase in the growing season (Kirilenko and Sedjo, 2007). Recent analysis using long-term, whole ecosystem measurements has found evidence to support the carbon dioxide fertilisation effect in the temperate and boreal forests of North America (Keenan *et al.*, 2013). Taken together, this shift in the productivity of trees and forests has significant implications for policy makers and global wood and forest social-ecological systems in general.

Another influence on the transition to cultivated wood production will be the availability of land and water. Past estimates have suggested that there are significant opportunities for expansion of wood cultivation estates, although not without challenges (Mola-Yudego *et al.*, 2017; Nilsson and Schopfhauser, 1995; Silva *et al.*, 2018). But while it is likely that increasing competition for land – resulting from globalisation, population increase and increasing prosperity – will continue (Lambin and Meyfroidt, 2011), as has already noted this is not just a competition for land (or other limited resources). These pressures also act to further incentivise wood cultivation into more concentrated forms, and wood production into more wood saving. This is happening in the shift to cellulose fibre production rather than solid wood production. Interestingly, this also suggests future possibilities for fibre cultivation from sources other than wood, such as through grass crops – including sugar-cane, sorghum and maize (Weijde *et al.*, 2013) and microbial sources (Keshk, 2014). These influences act on the wood cultivation systems with potentially unforeseeable consequences. They stress the complexity of wood and forest social-ecological systems and the immense challenges of working within these and the challenges of making predictions – predictions that are nevertheless key tools for policy makers.

The emergence of distinct realms of wood production and forest use is not necessarily a move against holistic or integrative approaches to land use. While wood production might become more focused geographically, its emergence into agricultural and other heavily modified landscapes will require new thinking about how wood production integrates within these landscapes. Landscape rather than forest thinking is now needed. Such approaches are emerging in the development of the 'landscape approach' to forestry, conservation and agricultural land development and management (Sayer *et al.*, 2013). This new landscape approach to land use planning and science integrates (subsumes?) forestry into a broader landscape thinking. Likewise, as forest ecosystem management is increasingly less focused on wood extraction and more focused on delivering a diverse range of non-wood values this will present separate challenges. New solutions will be required to successfully integrate these functions and demands into society and economics; in particular, finding ways for society to pay for the management of those values when the forests are not producing readily marketable wood products. While these challenges are considerable, strong forces are pushing the transformation of the wood and forest social-ecological system to these new post-forestry basins. This in turn provides an opportunity to imagine new sustainabilities for the world's forest ecosystem and wood production systems.

However, taking these opportunities also relies on the extent to which the transformational change into new system states is allowed or resisted. From the above there are still questions as to whether the stewardship forestry basin has more scope for adaptation, or if the time for adaptation has passed and effort should be redirected to transformation. Many detailed inquiries, and a sizable institutional effort, signify commitment to adaptation as a response to the emerging contradictions in the stewardship forestry basin. Still, there are limits to adaptation and its attempt to maintain 'the function [and] structural identity … of that system' (Nelson *et al.*, 2007, p. 397). This will include the point at which the alignment of ideals and technology that define core basin characteristics becomes changed to the extent that maintenance of the basin becomes untenable. At this point transformation becomes the logical and possibly inevitable change process, and further adaptation can become an increasingly futile resistance to transformation. System resilience, then, is more likely to come from transformation. Failure to maintain cognisance of the mutability of ideals and how they underpin a basin can lead to myopic efforts at adaptation and failure to address the possibility that the basin itself is no longer an ideal state to be maintained. Understanding this process of system change will allow improvements in policy. In particular, it will allow recognition that these changes can be used to develop policy that incorporates allowance for likely paths of change, ideally before they reach crisis points (Young, 2010). It should also result in policy that better manages confrontation between change in state variables and inertia within systems, in the form, for example, of distributional coalitions and institutional resistance.

Taken together, the above factors suggest that as wood production and forest ecosystem management transform into new domains the central institutions will have to alter. For stewardship forestry this suggests that its institutions, including that of the role of 'forester', will decline along with the method of wood production it 'owns' – the broad scale stewardship of natural and semi-natural forests. As new basins emerge natural resource managers might increasingly take the forest ecosystem management role while agricultural land managers and agronomists take on the cultivation of wood crops. While resource use systems in certain basins or states might be seen to be failing in resilience (for example when rural forestry industries decline or collapse), at other scales and in other locations new systems and opportunities emerge. In this sense, then, on a larger scale the resource use system displays considerable resilience through transformation to new conditions – wood production is resilient *through* its transformation.

Attempts to sustain the nexus of wood production and forest ecosystems that underpin the basin and to resolve key tensions in the wood and forest social-ecological system, by limiting analysis and adaptation to the domain of forests alone, will continue to struggle. Consideration needs to be given to *post-forestry* thinking and approaches as mandated by the institutional and geographic divergence of wood production and forest ecosystem management. These signify major changes in wood and forest social-ecological systems that are of vital import in the development of policy and management regimes for forests, agricultural lands and the societies that interact with them. From the perspective of the stewardship forestry basin and its institutions of forestry this raises the challenge of confronting *post-forestry* conditions – of considering the possibilities of transformative change.

Notes

1 Natural forest wood here includes semi-natural forest wood sources. Estimates for cultivated wood is average of high and low plantation scenarios from the chapter *Global analysis of trends in wood sourcing* and total wood from FAOSTAT (2016).
2 He cites Binkley (1997) for the certified natural forest wood production figure.
3 Exploitative removal of the highest grades of timber from a forest generally without regard for the longer term silvicultural needs of the forest.

References

Andersson, K., Lawrence, D., Zavaleta, J. and Guariguata, M.R. (2016). More trees, more poverty? The socioeconomic effects of tree plantations in Chile, 2001–2011. *Environmental Management*, 57, 123–136.

Angelsen, A. and Kaimowitz, D. (2001). Introduction: The role of agricultural technologies in tropical deforestation. In: Angelsen, A. and Kaimowitz, D. (eds.) *Agricultural technologies and tropical deforestation*. Oxon: CABI.

Batra, P. and Pirard, R. (2015). Is a typology for planted forests feasible, or even relevant? *CIFOR Infobrief*. Bogor, Indonesia: Center for International Forestry Research.

Bauhus, J., Van Der Meer, P. and Kanninen, M. (2010). *Ecosystem goods and services from plantation forests*. Oxon: Earthscan.

Bekessy, S.A., Wintle, B. A., Gordon, A., Fox, J.C., Chisholm, R., Brown, B., Regan, T., Mooney, N., Read, S.M. and Burgman, M.A. (2009). Modelling human impacts on the Tasmanian wedge-tailed eagle (*Aquila audax fleayi*). *Biological Conservation*, 142, 2438–2448.

Bengston, D.N., Xu, G. and Fan, D.P. (2001). Attitudes toward ecosystem management in the United States, 1992–1998. *Society & Natural Resources*, 14, 471–487.

Binkley, C.S. (1997). Preserving nature through intensive plantation forestry: the case for forestland allocation with illustrations from British Columbia. *The Forestry Chronicle*, 73, 553–559.

Binkley, C.S. (2003). Forestry in the long sweep of history. In: Teeter, L.D., Cashore, B.W. and Zhang, D. (eds.) *Forest policy for private forestry: global and regional challenges*. Oxon: CABI.

BMU, (Federal Ministry for the Environment, Nature Conservation and Nuclear Safety) (2007). National strategy on biological diversity. In: BMU (ed.). Berlin: BMU.

Brockerhoff, E.G., Jactel, H., Parrotta, J.A., Quine, C.P. and Sayer, J. (2008). Plantation forests and biodiversity: oxymoron or opportunity? *Biodiversity and Conservation*, 17, 925–951.

Buckingham, K.C., Wu, L. and Lou, Y. (2014). Can't see the (bamboo) forest for the trees: Examining bamboo's fit within international forestry institutions. *Ambio*, 43, 770–778.

Buongiorno, J., Zhu, S., Raunikar, R. and Prestemon, J.P. (2012). Outlook to 2060 for world forests and forest industries: a technical document supporting the Forest Service 2010 RPA assessment. Ashville. NC: United States Department of Agriculture.

Carle, J. and Holmgren, P. (2003). Definitions related to planted forests. *UNFF intersessional expert meeting international steering group on "the role of planted forests in sustainable forest management. Maximising planted forests' contribution to SFM"*. Wellington.

D'amato, D., Rekola, M., Wan, M., Cai, D. and Toppinen, A. (2017). Effects of industrial plantations on ecosystem services and livelihoods: perspectives of rural communities in China. *Land Use Policy*, 63, 266–278.

Daily, G.C., Polasky, S., Goldstein, J., Kareiva, P.M., Mooney, H.A., Pejchar, L., Ricketts, T.H., Salzman, J. and Shallenberger, R. (2009). Ecosystem services in decision making: time to deliver. *Frontiers in Ecology and the Environment*, 7, 21–28.

Elias, P., Boucher, D., Cummings, C., Goodman, L., May-Tobin, C. and Mulik, K. (2012). Wood for good: solutions for deforestation-free wood products. Cambridge, MA: Union of Concerned Scientists.

FAO (2015). *Global forest resources assessment 2015: how are the world's forests changing?* Rome: Food and Agriculture Organization of the United Nations.

FAOSTAT (2016). *Statistical databases*. Rome: Food and Agriculture Organization of the United Nations.

Fenning, T.M. and Gershenzon, J. (2002). Where will the wood come from? Plantation forests and the role of biotechnology. *Trends in Biotechnology*, 20, 291–296.

Ferguson, I. (2012). Future forestry employment and education. *Australian Forestry*, 75, 192–199.

Fischer, J., Batáry, P., Bawa, K. S., Brussaard, L., Chappell, M. J., Clough, Y., Daily, G.C., Dorrough, J., Hartel, T. and Jackson, L.E. (2011). Conservation: limits of land sparing. *Science*, 334, 593.

Fox, T.R. (2000). Sustained productivity in intensively managed forest plantations. *Forest Ecology and Management*, 138, 187–202.

Giddens, A. (1990). *The consequences of modernity*. Cambridge: Polity Press.

Glück, P. (1987). Social values in forestry. *Ambio: A Journal of the Human Environment*, 16, 158–160.

Greening, L.A., Greene, D.L. and Difiglio, C. (2000). Energy efficiency and consumption – the rebound effect – a survey. *Energy Policy*, 28, 389–401.

Hyde, W.F. *The global economics of forestry* (2012). New York: Resources for the Future.

Ingalls, M.L., Meyfroidt, P., To, P.X., Kenney-Lazar, M. and Epprecht, M. (2018). The transboundary displacement of deforestation under REDD+: Problematic intersections between the trade of forest-risk commodities and land grabbing in the Mekong region. *Global environmental change*, 50, 255–267.

Jevons, W.S. (1865). *The coal question: an inquiry concerning the progress of the nation, and the probable exhaustion of the coal-mines*. London: Macmillan.

Jürgensen, C., Kollert, W. and Lebedys, A. (2014). Assessment of industrial roundwood production from planted forests. *FAO Planted Forests and Trees Working Paper*. Rome: Food and Agriculture Organization of the United Nations.

Kant, S. (2000). A dynamic approach to forest regimes in developing economies. *Ecological Economics*, 32, 287–300.

Keenan, T.F., Hollinger, D.Y., Bohrer, G., Dragoni, D., Munger, J.W., Schmid, H.P. and Richardson, A.D. (2013). Increase in forest water-use efficiency as atmospheric carbon dioxide concentrations rise. *Nature*, 499, 324–327.

Kennedy, J.J., Thomas, J.W. and Glueck, P. (2001). Evolving forestry and rural development beliefs at midpoint and close of the 20th century. *Forest Policy and Economics*, 3, 81–95.

Keshk, S. (2014). Bacterial cellulose production and its industrial applications. *Journal of Bioprocessing & Biotechniques*, 4.

Kirilenko, A.P. and Sedjo, R.A. (2007). Climate change impacts on forestry. *Proceedings of the National Academy of Sciences*, 104, 19697–19702.

Lambin, E.F. and Meyfroidt, P. (2011). Global land use change, economic globalization, and the looming land scarcity. *Proceedings of the National Academy of Sciences*, 108, 3465–3472.

Lane, M.B. and McDonald, G. (2002). Towards a general model of forest management through time: evidence from Australia, USA, and Canada. *Land Use Policy*, 19, 193–206.

Leakey, A.D.B., Ainsworth, E.A., Bernacchi, C.J., Rogers, A., Long, S.P. and Ort, D.R. (2009). Elevated CO2 effects on plant carbon, nitrogen, and water relations: six important lessons from FACE. *Journal of Experimental Botany*, 60, 2859–2876.

Lebedys, A. and Li, Y. (2014). Contribution of the forestry sector to national economies, 1990–2011. *Finance Working Paper*. Rome: Food and Agricultural Organisation of the United Nations.

Leslie, A. (2005). What will we want from the forests? *ITTO Tropical Forest Update*, 15, 14–16.

Lindenmayer, D. (2016). Halting natural resource depletion: engaging with economic and political power. *The Economic and Labour Relations Review*, 28, 41–56.

Lindenmayer, D.B., Laurance, W.F. and Franklin, J.F. (2012). Global decline in large old trees. *Science*, 338, 1305–1306.

Lunney, D. (1987). Effects of logging, fire and drought on possums and gliders in the coastal forests near Bega, NSW. *Wildlife Research*, 14, 263–274.

Meyerhoff, J., Angeli, D. and Hartje, V. (2012). Valuing the benefits of implementing a national strategy on biological diversity – the case of Germany. *Environmental Science & Policy*, 23, 109–119.

Milder, J.C., Scherr, S.J. and Bracer, C. (2010). Trends and future potential of payment for ecosystem services to alleviate rural poverty in developing countries. *Ecology and Society*, *15*, 4. [online] www.ecologyandsociety.org/vol. 15/iss2/art4/.

Millennium Ecosystem Assessment (2005). *Ecosystems and human well-being: synthesis*. Washington, DC: Island Press.

Moeliono, M., Pham, T.T., Bong, I.W., Wong, G.Y. and Brockhaus, M. (2017). Social Forestry – why and for whom? *Forest and Society*, 1, 1–20.

Mola-Yudego, B., Arevalo, J., Diaz-Yanez, O., Dimitriou, I., Haapala, A., Ferraz Filho, A.C., Selkimäki, M. and Valbuena, R. (2017). Wood biomass potentials for energy in northern Europe: Forest or plantations? *Biomass and Bioenergy*, 106, 95–103.

Munck af Rosenschöld, J., Rozema, J.G. and Frye-Levine, L.A. (2014). Institutional inertia and climate change: a review of the new institutionalist literature. *Wiley Interdisciplinary Reviews: Climate Change*, 5, 639–648.

Nelson, D.R., Adger, W.N. and Brown, K. (2007). Adaptation to environmental change: contributions of a resilience framework. *Annual review of Environment and Resources*, 32, 395.

Nilsson, S. and Schopfhauser, W. (1995). The carbon-sequestration potential of a global afforestation program. *Climatic change*, 30, 267–293.

O'Brien, M. and Bringezu, S. (2017). What is a sustainable level of timber consumption in the EU: toward global and EU benchmarks for sustainable forest use. *Sustainability*, 9, 812.

Oliver, C.D. and Mesznik, R. (2006). Investing in forestry: opportunities and pitfalls of intensive plantations and other alternatives. *Journal of Sustainable Forestry*, *21*, 97–111.

Paquette, A. and Messier, C. (2009). The role of plantations in managing the world's forests in the Anthropocene. *Frontiers in Ecology and the Environment*, 8, 27–34.

Penna, I. (2010). Understanding the FAOs' wood supply from planted forests' projections. Ballarat, Australia: Centre for Environmental Management, University of Ballarat.

Phalan, B., Onial, M., Balmford, A. and Green, R.E. (2011). Reconciling food production and biodiversity conservation: land sharing and land sparing compared. *Science*, 333, 1289–1291.

Pierson, P. (2000). The limits of design: explaining institutional origins and change. *Governance*, 13, 475–499.

Pirard, R., Dal Secco, L. and Warman, R. (2016). Do timber plantations contribute to forest conservation? *Environmental Science & Policy*, 57, 122–130.

Pirard, R., Petit, H. and Baral, H. (2017). Local impacts of industrial tree plantations: An empirical analysis in Indonesia across plantation types. *Land Use Policy*, 60, 242–253.

Porter-Bolland, L., Ellis, E.A., Guariguata, M.R., Ruiz-Mallén, I., Negrete-Yankelevich, S. and Reyes-García, V. (2012). Community managed forests and forest protected areas: An assessment of their conservation effectiveness across the tropics. *Forest ecology and management*, 268, 6–17.

Rasker, R. and Hackman, A. (1996). Economic development and the conservation of large carnivores. *Conservation Biology*, 10, 991–1002.

Rist, L. and Moen, J. (2013). Sustainability in forest management and a new role for resilience thinking. *Forest Ecology and Management*, 310, 416–427.

Roessiger, J., Griess, V.C. and Knoke, T. (2011). May risk aversion lead to near-natural forestry? A simulation study. *Forestry*, 84, 527–537.

Sayer, J., Sunderland, T., Ghazoul, J., Pfund, J.-L., Sheil, D., Meijaard, E., Venter, M., Boedhihartono, A. K., Day, M. and Garcia, C. (2013). Ten principles for a landscape approach to reconciling agriculture, conservation, and other competing land uses. *Proceedings of the national academy of sciences*, 110, 8349–8356.

Sedjo, R.A. and Botkin, D. (1997). Using forest plantations to spare natural forests. *Environment: Science and Policy for Sustainable Development*, 39, 14–30.

Silva, L.N., Freer-Smith, P. and Madsen, P. (2018). Production, restoration, mitigation: a new generation of plantations. *New Forests*, 1–16.

Taub, D. (2010). Effects of rising atmospheric concentrations of carbon dioxide on plants. *Nature Education Knowledge*, 1, 21.

Weijde, T.V.D., Alvim Kamei, C.L., Torres, A.F., Vermerris, W., Dolstra, O., Visser, R.G.F. and Trindade, L.M. (2013). The potential of C4 grasses for cellulosic biofuel production. *Frontiers in Plant Science*, 4, 107.

Wiersum, K.F. (1995). 200 years of sustainability in forestry: lessons from history. *Environmental Management*, 19, 321–329.

Young, O.R. (2010). Institutional dynamics: resilience, vulnerability and adaptation in environmental and resource regimes. *Global Environmental Change*, 20, 378–385.

Zhang, Y. (2001). Economics of transaction costs saving forestry. *Ecological Economics*, 36, 197–204.

9 Resilience, innovation and sustainability in resource use

Forest ecosystems and timber are important natural resources that are fundamental to many societies. The social-ecological systems of these resources have been subject to globalisation, modernity and industrialisation with rapidly evolving social and technological changes profoundly influencing system development. It is in this milieu that the stewardship forestry system was forged with its wood/forest nexus and which is being subsequently riven into distinctive wood cultivation and forest ecosystem management systems. This divergence is a source of system change, political contestation and policy challenge. The last chapter considered the implications of these changes for wood and forest management social-ecological systems, in particular for stewardship forestry, as these systems emerge into post-forestry states. Here we will consider the implications of these changes for resource use systems more generally. First, and staying mostly with wood and forests, there is a brief review of some considerations or caveats in using the resilience model that should be kept in mind when applying them to large scale resource use social-ecological systems. Then the chapter looks at how innovation and sustainability interact, constantly shifting the sustainability equation, and in doing so also considers how changes in wood and forest social-ecological systems fit with theories of resource use change through industrialisation. Finally, the chapter and the book conclude by considering how the social in social-ecological systems effects system dynamics. It looks at: how late modernity has come to see nature as bound within society rather than as an external frontier; how the social can act as a source of system inertia as actors constantly steer the system to maintain stasis; how this raises tensions within the system as competing discourse coalitions argue differing change responses; and, finally, it considers how resource use futures are imagined in social-ecological systems where those researching and making decisions are bound within the system.

Resilience thinking in resource use social-ecological systems

Resilience thinking is a useful tool in understanding sustainability and change in resource use social-ecological systems – hence its use here. But this analysis has also highlighted some important considerations to keep in mind when using it. Some of these were noted in the initial discussion about this model but are reiterated here in the context of what was learnt from the case studies.

The first consideration is that the model suffers the critical limitation of all models that try to describe complex and diverse processes – there is often a gap between the generalised and the specific. The model needs to be treated as a framework for analysis, but its limits kept in mind as it deals with the specific. If treated this way the general model can serve to structure comparison generating system-wide insight; while being cognisant of the specific that does not fit so neatly can provide insight into the case being considered. The model provides a frame within which to consider a country or region with a distinctive wood and forest social-ecological system, but, nevertheless, it is clear that all nations or regions will involve unique circumstances. Caution and nuance are needed when applying the model to individual countries or regions. Particularly striking in the case studies here has been the difference between the conceptual 'flip' between states described in the general resilience thinking model and the nature of change exhibited in the case study countries. Rather than a flip, change between the basins described occurs as an iterative, lengthy and sometimes messy process with aspects of the before and after basins often coexisting to various degrees.

A second consideration is that the *social*, within the social-ecological, can contribute to basin stability. Social processes can act to create an internal equilibrium – a 'gravity' that stabilises a basin in the face of fluctuations and pressures in system conditions. Classic resilience analysis has tended to see ecological stability as the basis for basin gravity in the face of changing social conditions. What we see here is the social can also act to give a system resilience. Examples of this, described before, include the pull of path dependence and sunk costs of Australia's national plantation investments, the demands of vested interests and distributional coalitions in Indonesia, and the draw of discourse coalitions and professional institutions in New Zealand.

The third consideration for resilience thinking highlighted by this model and its case studies is the influence of the historical context on system conditions. The model worked well to describe the stewardship forestry basin as it emerged in the twentieth century on a global scale and in numerous countries around the world. As has been noted, the experience in Europe and the development of stewardship forestry there does not align so well with the model in three critical regards. One, was that the emergence of stewardship forestry in Europe was integrated with the emergence of wood cultivation – there were little extant natural forest left to manage at the time of stewardship forestry's emergence – new ones needed to be planted. Two, was the European formation of stewardship forestry prior to the ecological impulse of the late twentieth century. While the setting aside of forest for non-wood values had occurred before the twentieth century, the examples are limited and generally small in scale (for example, sacred forest groves or royal hunting grounds). The extent of the valuing of other ecosystem services from forests that emerged in the twentieth century was unprecedented in the extent to which it created pressure on the internal logics and institutions of wood and forest social-ecological systems. Despite these differences, now, as noted earlier, the wood and forest social-ecological systems of Europe are subject to these same pressures from the rise of social valuation of the ecological within forests and displaying some signs of shifts to post-forestry conditions. The third critical regard in relation to Europe's distinctive historical context was the subsequent and unique role of European colonial, technological and cultural imperialism that spread this particular brand of stewardship forestry globally. The stewardship forestry basin that emerged globally was informed by the silvicultural thinking of these European origins. The rapid spread and uptake of the 'technologies' of forestry, along with other features of modernity – the nation state, science and capitalism – that developed in Europe has in many ways been a crucial panarchy feature for the balance of the world's wood and forest social-ecological systems.

While Europe had these globally significant historical distinctions, the effects of historical conditions, both as internal and external conditions, can also be seen in the case studies. In these there was also an interplay of history within country and panarchy influences. For example, when European colonisation of New Zealand commenced its forests were already significantly depleted. This depletion then accelerated. When the institutions of stewardship forestry developed, they did so quickly as they transposed the ideas of European and colonial stewardship forestry to the remaining indigenous forests. In spite of that, it took some time for the understanding of the slow growth and regeneration of the unique forests of New Zealand to be fully realised. In the end, stewardship forestry in New Zealand only managed to preside over the slow exhaustion of indigenous forest wood production. By comparison, Australia has been more successful at managing the stewardship forestry phase of its natural forest wood extraction. In part, this is because there were more extensive, and productive, natural forests to include in the stewardship forestry estate to begin with. Even more striking is the case of Indonesia where the arrival of stewardship forestry came at a point where the extant natural forests were still largely in place and yet to be exploited. In all three cases the implementation of colonial stewardship forestry got under way in the early twentieth century, but in each case there were unique local conditions. These differences highlight the significance of both local distinctiveness and historical context in understanding how this model is expressed in unique places.

These are useful considerations for using the model for wood and forest social-ecological systems, but also for analysing resource use social-ecological systems more generally. Social-ecological systems will not neatly match the model due to much diversity in system conditions, including historical context, and they do not necessarily make neat switches between states. However, it is clear that the model provides a useful conceptual and analytical framework for resource use social-ecological systems. Resilience concepts help when describing and analysing complex sets of interacting phenomena, particularly when undertaking multidisciplinary and cross-disciplinary analysis. Furthermore, while resilience, adaptation and transformation are described as three different forms of response to system change and pressures, what is also clear from the above is that these three forms of change can be seen in action at the same time. These different forms of change will offer different outcomes for different actors and different parts of social-ecological systems, and can lead to contestation over the ideal change response. This begs the questions: for whom is the system best kept resilient, adapted or transformed? These questions become fundamental to resolving issues of sustainability, resilience and the future growth or maintenance of resource use social-ecological systems.

Innovation, sustainability and industrialization

In September 2016, the New Zealand Government announced seed funding for research into 3D printing using live tree wood cells. The research will look at how to trigger the 3D printed cells wooden structure after printing a specified shape, turning the printed form to wood – all 'without the need for the destructive harvesting of trees' (University of Canterbury, 2016). Printing wooden products from cultured tree cells is a logical (if considerable) further evolution in the process of increasingly intensive and technological wood production. It is consistent with a pattern of shifting wood production from solid timber to reconstituted wood fibre products and of shifting wood growing from extant forest exploitation to intensive fibre cultivation. It highlights the potential for innovation to act as a destabiliser to wood and forest social-ecological systems, but also to respond to system pressures such as growing demand and limited supply.

The process of induced innovation in wood production is consistent with the Boserup theory of agricultural production (espoused by the Danish economist, Ester Boserup). The theory states that technological innovation in agriculture comes about in response to increased demand resulting from population growth. This was proposed in opposition to a Malthusian perspective (Boserup, 1965) that growth in demand for agricultural output will outstrip capacity for the system to supply – innovation is limited and certainly not to be relied upon. Applied to wood (note that Malthus focused on food agriculture), the Malthusian perspective would have it that a growing population and demand for wood will lead to overharvest and consequent system collapse (there is certainly considerable evidence of wood resources being used in non-sustainable patterns, and some historical instances of societal collapse following this). However, as the Boserup theory suggests, population pressure (demand) on resource use social-ecological systems can respond to demand and supply mismatches through innovation rather than system collapse, decline or restriction. Notably, in complex systems there will be multiple places in the system where innovation can occur, continuously shifting the supply and demand balance. This capacity to innovate becomes critical for appraising the sustainability and resilience of a system.

The two views (Boserup's and Malthus') are often framed as opposing. And, as has been noted, there is evidence to support both processes at play in relation to wood. However, there is also the possibility that they are complementary. It could be useful to consider that Malthusian analysis acts as a feedback mechanism within the system – a critical alarm that the system is heading in the wrong direction. But, once this alarm is raised in the system then Boserup processes come into play, acting to change the very conditions that gave rise to the initial Malthusian analysis (for a similar idea of how the two theories can complement one another see Demont *et al.*, [2007]). System resilience and sustainability is then achieved, not through either process alone, but by their interaction and dual contributions.

While the theories of Boserup and Malthus respond to increased demand generated by population growth in industrialising societies, there are also demand and supply pressures induced in late-stage industrialisation. This can be seen in the interplay between innovation and demand responding to new social demands for ecological and recreational values from forests. Pressure to preserve forests for these values can act as a source of timber supply restriction, through regulatory harvest restrictions or prohibition of all wood production. In the same way that a population increase or growing wealth increases demand, these restrictions on supply can act to induce innovation in wood cultivation and wood product development. This is the Porter hypothesis – environmental regulation induces innovation and thus has a positive effect on economic competitiveness and system sustainability (Howarth, 2012; Porter, 1996). Restrictions on forest harvest arising from environmental regulation (through harvest restrictions and regulation) will generally contribute to a reduced harvest, which in turn raises costs and thus drives innovation. As noted above, much of this innovation in wood production is occurring through shifts to intensive wood cultivation and wood saving technologies. Wood production from natural forest ecosystems then gets reduced, both through regulatory restrictions and the resulting innovation creating new wood cultivation systems that shift supply out of the forests. The effect of increasing environmental regulation, then, is a positive impact on natural forests by furthering the wood sourcing transition – in line with the Porter hypothesis.

It is also worth noting that the pattern of change in wood and forest social-ecological systems in the face of industrialisation and growing demand outlined here is explained in a number of other models of environmental change, notably forest transition theory and the environmental Kuznets curve. Forest transition theory discusses a given jurisdiction's shift from net forest loss to net forest gain – it is a 'historical generalization' of these observed changes (Rudel *et al.*, 2010, p. 95). It posits that the extent of a country's forested area goes through a period of decline during early stages of industrialisation, which then reverses in later stages, with *transition* nations being those that have shifted from total forest loss to total forest gain (Meyfroidt and Lambin, 2011). However, it has been noted that in some cases this increase in forest occurs as a result of increased wood cultivation, especially industrial timber plantations (Farley, 2007), and the process of leakage of wood production from countries with growing forest estates, noted previously in the examples of Japan and Vietnam, mean the forest transition does not always deliver the ecological benefit implied by a growing forest estate.

This forest transition also shares conceptual ideas with the *environmental Kuznets curve*, with both postulating that environmental outcomes worsen during early industrialisation phases, but subsequently improve. The idea of the curve is an extrapolation of the economist Kuznets' theoretical work (Kuznets, 1955) which suggested that in its early phase industrialisation led to increasing wealth inequality, but that this trend then reversed as the industrialisation phase matured. Applied to environmental matters, the basic idea of the *environmental Kuznets curve* is that environmental degradation will increase in the early stages of industrialisation but decrease in the later stages (as with the forest transition theory). Recent reviews on the environmental Kuznets curve highlight the ambivalence of the empirical data in backing up the theory (Caviglia-Harris *et al.*, 2009; Chowdhury and Moran 2012; Kijima *et al.*, 2010), although some note that the data is stronger for certain types of environmental degradation, rather than others.

The forest transition idea and the environmental Kuznets curve describe a process of sustainability being achieved over the long run of industrialisation. The changes in wood and forest social-ecological system outlined in this book provide supportive evidence for these generalised processes occurring in relation to forests and wood. Carson (2010) notes how these positive transitions in response to early resource use social-ecological system unsustainability contradict earlier sustainability models which postulated that environmental impact simply grew with growth in population, affluence and technology (the IPAT model – Impact = Population/Affluence/Technology). These ideas were popular in environmental analyses such as those of the Club of Rome (Meadows *et al.*, 1972) and the Ehrlichs (Ehrlich and Ehrlich, 1972) which argued for the need to protect nature from these inherently destructive processes. In some ways this is the logic of the stewardship forestry period – the impulse to protect forests from rampant and unchecked growth. But this pattern, describing a trajectory toward ruin (given ever growing populations, affluence and technology), is in some critical ways contradicted by the evolution of intensive wood cultivation and new forest ecosystem values noted here. Either this suggests the technology influence is stronger than earlier pessimistic assessments of the IPAT model, or it could be explained by the absence of changes in behaviour and values being considered in the IPAT equation (Carson, 2010). Another way of looking at this is to consider changing *Technology* to *Innovation* so that the equation becomes IPAI. Here *Innovation* embraces not just technology in the narrow sense of the term but embraces the broader suite of social innovations occurring in response to resource use pressures. These social innovations could include: improved governance, policy, legal frameworks and enforcement; realigned power relationships, politics and distributional and discourse coalitions; or, economic and market adjustments including substitution effects and trade. Innovations in any or all of these areas, along with scientific and technological innovation, greatly raise the power and potential to modify the sustainability equation.

The ecomodernist position, which posits the '[d]ecoupling of human welfare from environmental impacts [through] a sustained commitment to technological progress and the continuing evolution of social, economic, and political institutions' (Asafu-Adjaye *et al.*, 2015, p. 29), aligns with the above theories and has been particularly assertive in this position. Yet, these theories of positive response to unsustainable resource use remain contentious. Notably, such positive societal response to the environmental pressure of industrialisation can be critiqued for a blind techno-optimism. There have been forceful calls for caution, particularly in the focus on applying technology in such a way (Grunwald, 2016). But equally, the interplay of new social institutions and technologies is a significant caution to the use of simple extrapolations of current systems and trends to imply unsustainability. Wackernagel *et al.* (2004) note that these processes make it misleading to take static descriptions of the balance between resources and humans from which estimates of resource limits are drawn. The complexity of this interplay and thus the limits of extrapolating predictions of unsustainability can be seen in the wood and forest social-ecological system example. These complex systems have many ways in which changing demand pressure can be addressed. For example, wood cost pressure can be addressed by increasing the use of land for more plantations, it can also be met through at least four other key routes noted here: increasing productivity of existing plantations; utilising degraded and underutilised cleared lands; increasing the productivity of wood utilisation; or, shifting demand to non-wood materials.

More broadly, for resource use systems in general, Wackernagel *et al.* (2004, p. 273) note four ways that human societies tend to avoid the population/resource overshoot common to populations of other species – overshoot being a lag between population decline or stabilisation when a limiting resource reaches its limits. First, much resource demand by humans is not required for direct bodily consumption, but, rather, indirectly in the creation of goods, giving considerably more time and scope for substitution and innovation. Second, trade acts to dramatically increase the 'catchment' from which resources can be drawn. The third and fourth are both technological, the third being the ability to manage and manipulate ecosystems to dramatically escalate their productivity of wanted resources (as exemplified by aquaculture, agriculture and wood cultivation), and then there is technology's capacity to develop substitution. An example of this was shifting from wood for fuel to fossil fuel, much as renewable energy sources are now starting to address the issue of limits to fossil fuels.

Overall, the above highlights numerous ways in which wood and forest social-ecological systems can respond to resource pressures. Consider the critique of expanding plantations taking agricultural land and thus reducing food supply, a Malthusian analysis. Such an assessment requires an assumption that all other aspects of the food production system stay constant. They do not. As noted above there are many ways these complex systems can adapt to such pressures. Taken together, these complexities highlight the challenges of sustainability analysis on systems with so many possible ways for responsive change to occur.

The need to consider these complexities in how systems respond to demand and supply pressure requires multi and cross disciplinarity in both research and in policy development. Further, the interplay between various resource use systems, as exampled by wood and food above, commends a landscape wide approach in facing these sustainability conundrums. Such a landscape approach is demanded by the intense cross sectoral interactions across the broader landscape where wood cultivation interconnects with systems of agroforestry, agriculture, industrial development, urban expansion and other land uses.

Sustainability, inertia and reflexivity

The example here for timber and forests is one of many resource use social-ecological systems that evolve from systems extracting from the wild to cultivation and/or recycled systems. It can be seen in wild-ocean fishing shifting to aquaculture, or ore mining changing to the recycling of metals such as lead, aluminium or copper. For all such extractive systems, following a receding resource frontier can only be sustained for so long. It is not sustainable – no wild system has endless frontiers – and yet the society that is dependent on this process will have strong social systems acting to continue deriving the benefits from this process, often leading to a crisis in the system. The social-ecological system ultimately needs to shift from being one that extracts from (and possibly dumps into) an outside frontier to one that incorporates the cultivation and recycling of the resource within. The resource use social-ecological system becomes conceptually closer to being self-contained and self-sustaining than systems acting on frontiers.

This enclosing of the biophysical dimension of the resource use social-ecological system within itself is mirrored in late-stage modernity. The emergence of reflexive modernisation includes a shift of nature, previously seen as outside of society (the frontier from which resources were extracted), becoming internal (or integral) to society (Beck *et al.*, 1994). This perception in which society embraces (or integrates with) nature leads to new logics. Society is no longer extracting from, or dumping in, the 'other'. Nature becomes not something outside the society from which to draw resource as needed, but is embraced within. Now the extraction, or dumping, occurs within. Amongst other things, this changes how nature is understood and valued. It calls forth the need for a new ethos in resource stewardship.

Given the potential for a systemic crisis that the end states of frontier extraction will generate, ideally such a crisis will be averted by recognising it with sufficient time to make necessary changes. An analogy is that of a large ship on a trajectory (non-sustainable resource use) that will bring it onto a large rocky island (natural resource exhaustion and social-ecological system crisis) (Figure 9.1). There are many points at which the ship can start turning away to avoid the rock. But the longer it is left the more effort it will take to turn the ship (by analogy, the more disruptive it will be to the social-ecological system).

Systemic social inertia in the social-ecological system surrounding resource uses can be substantial, as noted previously, especially social (including economic and political) inertia. Actors within the system will act to maintain the current trajectory, while others will be calling out the impending crash with the rock and the need to change course. Those wishing to maintain the current course can be comprised of powerful distributional coalitions that have coalesced around the maintenance of a system even when engaged in a non-sustainable resource use trajectory. This is a key driver to the often intense political nature of conflicts within these systems. Ossified social relations and structures that support a given non-sustainable trajectory can respond slowly to their decreasing tenability in the face of increasingly obvious limits to certain patterns of resource use. Such debates about natural resource use systems can become intensely polarised and intractable (Dunlap and York, 2012; Guber, 2013; Lucas and Warman, 2018).

Large and complex societies have a momentum that makes change hard and forces them to continue with unsustainable rates of exploitation (Vogt *et al.*, 2007). Sub-groups that are mostly dependent on the resource are likely to be unaware of emerging limitations. They will continue to act on the belief there are no constraints upon their way of life. In the case of forests, '[t]his

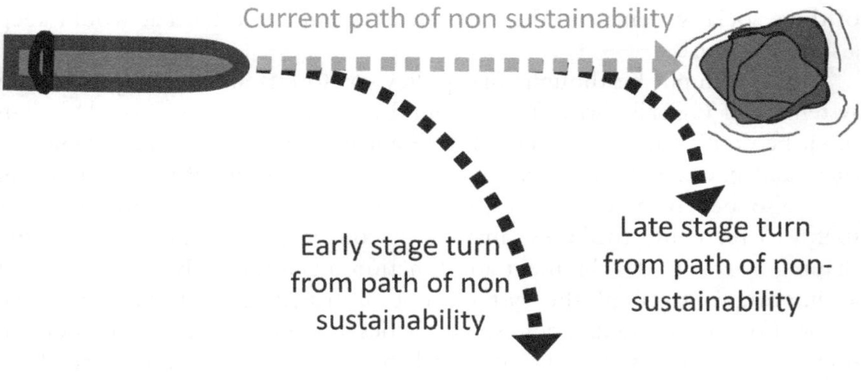

Figure 9.1 Representation of a ship (a social-ecological system) changing direction (transforming) to avoid running ashore on a rocky island (catastrophic collapse of system).

widely held view encourages the exploitation of forests and is a view that resurfaces throughout the history of global societies and forests' (Vogt *et al.*, 2007, p. 6). This can stifle effective change to social-ecological systems on multiple levels. Using the analogy above, this constitutes significant inertia in the system, analogous to a large ship. Changing direction in the face of such inertia takes considerable time and effort.

We have already touched on the extent of the social networks that emerged within the stewardship forestry basin. The central profession of the stewardship forestry basin, foresters, form part of an extensive discourse coalition that will include regional communities, timber companies, labour organisations and civil society groups acting in support of a status quo. Identity, along with ideology and political differences, will act to strengthen these discourse coalitions. They will tend to act together to maintain the system in a state that benefits them and actively resist change. In Tasmania, Australia, as one example, such a discourse coalition developed in the stewardship forestry basin acting to maintain the status quo in natural forest stewardship forestry (Lucas and Warman, 2018). However, as the late twentieth century environmental consciousness developed, there was also the formation of an opposing discourse coalition that proposed system state change, particularly in order to advance non-wood value protection of forest ecosystems subject to stewardship forestry management. These opposing discourse coalitions engaged in intense conflict over decades. Notably, there is evidence that the polarised conflict persisted from previous resource conflicts in the state (over dams and hydroelectricity) and into subsequent conflict (over climate change) (Lucas and Warman, 2018). The persistence of these opposing discourse coalitions across multiple resource conflicts have been called 'ruts' (Lucas and Warman, 2018), while the persistence of collective commitments to unsustainable resource use patterns, analogous to personal addictive behaviour, have been called 'social traps' or 'societal addictions' (Costanza *et al.*, 2017). These notions highlight the perseverance of these social systems and the strength of systemic inertia that would keep a ship on course.

While it might be thought that policy for forests and wood production systems will logically arise from a sound assessment of the facts, there can often be conflicting logics behind different identities, values and ideologies that challenge this. It is not enough to focus on 'matters of fact' – attention must also be given to 'matters of concern' (Latour, 2004). Further, it is useful to recognise that these matters of concern, as central factors in the shaping of system development and function, need themselves to be treated as important 'facts' of the system, not dismissed as irrationalities to be ignored or surmounted. These subjectivities are key features of the social in social-ecological systems and are fundamental given their power to drive system evolutions.

The implications of the clash of sustainabilities, and the obvious tensions in the use of the term *sustainability*, leads Colander (2005) to ask: Sustainability of what? For whom? In the case of stewardship forestry, institutions have been built on the logic of sustainability. From the perspective of the forest stewardship advocates who came to prominence in the nineteenth century, the definition of sustainability related to supply of wood and maintenance of forest cover to supply this wood. This changed over time as the non-wood qualities of forests became increasingly valued. Thus, sustainability as the capacity of a system to provide certain goods/services/values over a long period, is itself subject to the changing values defining sustainability (Colander's 'what?'). Defining what is sustainable (what is optimal) when the conditions of the social-ecological system (including social values) are evolving is a challenge. Indeed, resilience thinking theorists point out '[t]here is no such thing as an optimal state of a dynamic system' (Walker *et al.*, 2006, p. 141). The pursuit of sustainability, a system in some form of steady state, is confounded by the shifting of the very values for which sustainability is sought.

So, is sustainability possible? Achieving sustainability requires confronting social systems with strong inbuilt resistance to certain change, while the nature of sustainability is itself being continually reinvented (and contested). Despite these challenges, the analysis here of the response of societies to non-sustainable resource use trajectories – in this case wood and forest social-ecological systems – shows that such social-ecological systems do respond to unsustainability. In a sense they must, the logic of the non-sustainable trajectory is simple: it dictates the system will have to undergo change, if not planned, then saved by luck, or perhaps more likely, forged in crisis and contestation.

And while we are focused here on the contestation over sustainability of what and for whom, we also need to remember the earlier discussion about how innovation changes sustainability equations. Going back to the analogy of the system heading towards the rock, what effect does innovation have in the system? If, as theorised, innovation occurs in response to demand and supply pressures then it could follow that such innovation will increase as the pressure increases. To use the ship analogy, the pressure to change course becomes stronger as the ship approaches the rock. Indeed, such a response could be seen as a form of inbuilt failsafe for systems – a key element of long term resilience even as the system changes through the transformative power of innovation.

One of the hallmarks of reflexive modernity is the realisation that there are multiple futures, including different sustainabilities, for resource use social-ecological systems. A 'true' sustainability is not found by looking from outside society, but is determined within. This is understood in the theory of the risk society. Beck *et al.* (1994) note that the current concern with environmental risk is a consequence of modern efforts to conquer nature and overcome natural hazards. Beck labels debates over how to respond to tensions in resource use within the social-ecological system as 'risk conflicts', in which the debates are about different definitions of the risk (Beck, 1992). 'The success of the modern project, and the resulting human dominance over nature has led to the replacement of natural hazards by risks' (Lucas and Warman, 2018). These risks are described as 'manufactured uncertainties' – artefacts which are internal to society rather than external (Beck, 1992; Giddens, 1994). These manufactured uncertainties are integral to the contested sustainabilities noted above. They are generated within a social system that has integrated into its biophysical systems. And this social system that generates them must arbitrate on them, alone.

This internal nature of defining risk and sustainability in the risk society creates a fundamentally different proposition for how these risks are addressed. The use of natural resources, the manufacture of risk and the choosing of sustainabilities have become integrated within the social system. This puts these risks outside the scope of objective science alone to calculate, because they are not objective (Beck, 2016). Rather, they are constituted socially and in response to social-ecological systems that have complex feedback systems and unknown outcomes and future evolutions. In addition, the multiple actors and various discourse coalitions in the wood and forest social-ecological system all assert legitimacy and their own knowledge (Lucas and Warman, 2018) forming a network of interconnected individual and systemic interpretations of risk (Beck, 1992).

Of this modern era, Giddens (1990, p. 39) says '[we] are abroad in a world which is thoroughly constituted through reflexively applied knowledge, but where at the same time we can never be sure that any given element of that knowledge will not be revised.' In realising this, the fallacy that science and experts can rationally solve the problems of resource use social-ecological systems becomes apparent (Beck *et al.*, 2003). Indeed, scientific arguments and 'facts' will be invoked to justify quite divergent positions on the future of wood and forest social-ecological systems. These 'facts' can then masquerade as 'objective' arguments that serve only to mask the political, social and ideological motivations of those arguing (Corner, 2012; Kahan *et al.*, 2011). This is an important conceptual frame through which to view debates about wood and forest social-ecological systems. Debates over the use of forests, wood production, the possibility of sustainable forest management or the merits of land sharing versus land sparing approaches all occur within this space of reflexive modernity.

The research and writing of this book itself occurs within this reflexive modernity. As an exercise that has straddled the natural and social sciences it should be noted that this book's analysis entails a significant complication. Giddens (1990, p. 16) notes the peculiarity of social science which 'spirals in and out of the universe of social life, reconstructing both itself and that universe as an integral part of that process'. Unlike the natural science (of forestry, for example) which seeks to grow knowledge on a 'parallel track' to the object of its study, the social science can never truly gain that independence (Giddens, 1990). In this sense a wood and forest social-ecological system will always bind its analyst and researcher within itself, responding and evolving in response to their findings. Research of the wood and forest social-ecological system is also part of the system – not a removed outside phenomenon.

Beck would have it that such reflexivity is part of the zeitgeist of the modern era, a reflexivity born by the emergent properties of the problems and choices, now found *within* an all-embracing society. Beck says that the reflexive modernisation society must confront itself. The problem is not out there, but rather within. The challenge, Beck says, becomes one of answering, 'how do we wish to live?' (Beck, 1992, p. 119). This is pertinent when we respond to the way we do things when we know that it is unlikely to be sustainable, though we have much invested in this way. The daunting complexity, not least due to the presence of humans in the system, results in much uncertainty. The determination of what is right or wrong is dependent on contexts that are dynamic and fickle. Maybe, then, it is a matter of making a choice, of asking: 'how *do we wish* to live?', rather than striving to understand the answer to 'how *should* we live?'

Along these lines, Srnicek and Williams (2015) argue that the future of social systems is determined by the imaginings of those who have gone before. They use the example of a marginal group of economists in the middle of the twentieth century who set about pushing a heterodox economic liberalism (thinkers such as Hayek and Friedman). By the end of the twentieth century these ideas would form the basis of a new globally dominant hegemonic neoliberal order. Srnicek and Williams argue that this was not preordained but rather the result of a successful imagining of a future, and communication of that imagining. For wood and forest social-ecological systems, indeed for all resource use social-ecological systems, this leads to the question: what future is to be imagined?

In deciding the future of wood and forest social-ecological systems the foresters and forest biometricians have long imagined ways of growing wood supply while reconciling increasing demand for non-wood values. Countervailing this is the imagining of conservation ecologists for an extensive area of forest ecosystems managed without wood production, often through a large conservation estate (IntAct, 2017; Nature Needs Half, 2017), and ambitious 'rewilding' initiatives (Fraser, 2009). As has been noted, these imaginings are much contested. This book contributes to this contestation and posits a way forward through the shifting of wood production from forest ecosystem extraction to cultivation – a new post-forestry condition. This work then contributes to a new imagining. The plausibility of this imagining is supported by the processes examined here.

If stewardship forestry is transforming into divergent spheres of agricultural style wood cultivation and forest ecosystem management largely free of wood production, then the logic in *forestry* – wood production – no longer holds. The wood/forest nexus is undone. In that, then, forestry ceases to exist. We move to a post-forestry condition in wood and forest social-ecological systems. Wood production comes out of the 'woods' and enters broader landscapes, while forest ecosystem management diversifies around multiple values, calling forth new institutions – the new landscape approaches of a post-forestry world.

References

Asafu-Adjaye, J., Blomquist, L., Brand, S., Brook, B., Defries, R., Ellis, E., Foreman, C., Keith, D., Lewis, M., Lynas, M., Nordhaus, T., Pielke, R.J., Pritzker, R., Joyashree, R., Sagoff, M., Shellenberger, M., Stone, R. and Teague, P. (2015). *An ecomodernist manifesto* [Online]. The Breakthrough Institute. Available: www.ecomodernism.org/s/An-Ecomodernist-Manifesto.pdf.

Beck, U. (1992). *Risk society: towards a new modernity*. London: Sage Publications.

Beck, U. (2016). *The metamorphosis of the world: How climate change is transforming our concept of the world*. Cambridge: Polity Press.

Beck, U., Bonss, W. and Lau, C. (2003). The theory of reflexive modernization problematic, hypotheses and research programme. *Theory, Culture & Society*, 20, 1–33.

Beck, U., Giddens, A. and Lash, S. (1994). *Reflexive modernization: politics, tradition and aesthetics in the modern social order*. Stanford, CA: Stanford University Press.

Boserup, E. (1965). *The conditions of agricultural growth: the economics of agrarian change under population pressure*. London: George Allen & Unwin Ltd.

Carson, R.T. (2010). The environmental Kuznets curve: seeking empirical regularity and theoretical structure. *Review of Environmental Economics and Policy*, 4, 3–23.

Caviglia-Harris, J.L., Chambers, D. and Kahn, J.R. (2009). Taking the "U" out of Kuznets: a comprehensive analysis of the EKC and environmental degradation. *Ecological Economics*, 68, 1149–1159.

Chowdhury, R.R. and Moran, E.F. (2012). Turning the curve: a critical review of Kuznets approaches. *Applied Geography*, 32, 3–11.

Colander, D. (2005). Complexity, muddling through, and sustainable forest management. In: Kant, S. and Berry, R.A. (eds.) *Economics, sustainability, and natural resources*. Dordrecht, The Netherlands: Springer.

Corner, A. (2012). Evaluating arguments about climate change. *Perspectives on scientific argumentation*. Dordrecht, The Netherlands: Springer.

Costanza, R., Atkins, P.W., Bolton, M., Cork, S., Grigg, N.J., Kasser, T. and Kubiszewski, I. (2017). Overcoming societal addictions: what can we learn from individual therapies? *Ecological Economics*, 131, 543–550.

Demont, M., Jouve, P., Stessens, J. and Tollens, E. (2007). Boserup versus Malthus revisited: Evolution of farming systems in northern Côte d'Ivoire. *Agricultural Systems*, 93, 215–228.

Dunlap, R.E. and York, R. (2012). The Globalization of Environmental Concern. In: Steinberg, P.F. and Vandeveer, S.D. (eds.) *Comparative Environmental Politics: Theory, Practice, and Prospects*. Cambridge, Massachusetts: MIT Press.

Ehrlich, P.R. and Ehrlich, A. H. (1972). *Population, resources, environment: issues in human ecology*. San Francisco, CA: WH Freeman.

Farley, K.A. (2007). Grasslands to tree plantations: forest transition in the Andes of Ecuador. *Annals of the Association of American Geographers*, 97, 755–771.

Fraser, C. (2009). *Rewilding the world: dispatches from the conservation revolution*. New York: Picador.

Giddens, A. (1990)._*The consequences of modernity*. Cambridge: Polity Press.

Giddens, A. (1994). *Beyond left and right: the future of radical politics*. Stanford, CA: Stanford University Press.

Grunwald, A. (2016). Diverging pathways to overcoming the environmental crisis: a critique of eco-modernism from a technology assessment perspective. *Journal of Cleaner Production*.

Guber, D.L. (2013). A cooling climate for change? Party polarization and the politics of global warming. *American Behavioral Scientist*, 57, 93–115.

Howarth, N. (2012). Clean energy technology and the role of non-carbon price-based policy: an evolutionary economics perspective. *European Planning Studies*, 20, 871–891.

IntAct. (2017). *IntAct: international action for primary forests* [Online]. Available: http://primaryforest.org/ [Accessed 17/01/2017].

Kahan, D.M., Jenkins-Smith, H. and Braman, D. (2011). Cultural cognition of scientific consensus. *Journal of Risk Research*, 14, 147–174.

Kijima, M., Nishide, K. and Ohyama, A. (2010). Economic models for the environmental Kuznets curve: a survey. *Journal of Economic Dynamics & Control*, 34, 1187–1201.

Kuznets, S. (1955). Economic growth and income inequality. *The American Economic Review*, 45, 1–28.

Latour, B. (2004). Why has critique run out of steam? From matters of fact to matters of concern. *Critical Inquiry*, 30, 225–248.

Lucas, C. and Warman, R. (2018). Disrupting polarized discourses: Can we get out of the ruts of environmental conflicts? *Environment and Planning C: Politics and Space*, 36, 987–1005.

Meadows, D. H., Meadows, D. L., Randers, J. and Behrens, W.W. (1972). *Club of Rome. The limits to growth; a report for the Club of Rome's project on the predicament of mankind*. New York: Universe Books.

Meyfroidt, P. and Lambin, E.F. (2011). Global forest transition: prospects for an end to deforestation. *Annual Review of Environment and Resources*, 36, 343–371.

Nature Needs Half. (2017). *Nature needs half* [Online]. Boulder, CO: Nature Needs Half. Available: www.natureneedshalf.org/ [Accessed 17/01/2017].

Porter, M. (1996). America's green strategy. *Business and the environment: a reader*, 33.

Rudel, T.K., Schneider, L. and Uriarte, M. (2010). Forest transitions: an introduction. *Land Use Policy*, 27, 95–97.

Srnicek, N. and Williams, A. (2015). *Inventing the future: postcapitalism and a world without work*. New York: Verso Books.

University of Canterbury. (2016). *UC researcher aims to 3D-print synthetic wood* [Online]. Christchurch, New Zealand: University of Canterbury. Available: www.comsdev.canterbury.ac.nz/rss/news/?articleId=1968 [Accessed 06/02/2017].

Vogt, K., Gara, R., Honea, J., Vogt, D., Patel-Weynand, T., Roads, P., Fanzeres, A., Sigurdardottir, R., Honea, J. and Andreu, M. (2007). Historical perceptions and uses of forests. In: Vogt, K.A., Honea, J., Vogt, D.J., Andreu, M., Edmonds, R., Sigurdardottir, R. and Patel-Weynand, T. (eds.) *Forests and society: sustainability and life cycles of forests in human landscapes*. Wallingford, UK: CAB International.

Wackernagel, M., Monfreda, C., Schulz, N.B., Erb, K.-H., Haberl, H. and Krausmann, F. (2004). Calculating national and global ecological footprint time series: resolving conceptual challenges. *Land use policy*, 21, 271–278.

Walker, B., Salt, D. and Reid, W. (2006). *Resilience thinking: sustaining ecosystems and people in a changing world*. Washington, DC: Island Press.

Index

Page numbers in **bold** denote tables, those in *italics* denote figures.

Acacia 28, 54
actor-network theory 2–3
actors 3, 47, 85, 111, 176, 181
adaptation 48, 52–53, 144–145, 162, 167–168, 176
adat see customary land practices
advantage, comparative 51, 79, 101
Africa 31, 32, 73–74
agricultural science 53
agriculture 143, 178, 179–180; cultivated wood production as 1, 6, 46, 108, 185; early 17–20; land clearing 15–16, 19, 33, 35, 50, 108; logic of 154, 156–156; *see also* deforestation
agroforestry 4, 6, 28, 31, 53, 74, 120, 155, 180
agronomy 147, 168; and tree cultivation 29, 40, 140; *see also* agricultural science
aid *see* development assistance
Amazon 18, 29, 57
Asia 28, 29, 31, 32; Southeast Asia 28, 38n3, 53, 123n5
Asian financial crisis 108, 115
attractors 47, 51, 162
Australia 57, 66, 73, 85–103, 160–161, 175; deforestation 33; forest policy 26–27; introduction 7–8; location 7; plantation conservation benefit 157–158; plantation policy 26–27, 38n6, 159; wood production 8, *9*

balance, in forestry 1, 25–26, 66, 179
bamboo 32, 162
banjir kap 113, 116
basins, of attraction 3, 46–58, 107–108, 122, 128–131, 137, 139, 143–146, 154, 167–168, 174–175
big bang, in Indonesia 110, 117, 120–122, 128
biodiversity 27, 36–37, 50, 55, 66, 79–80, 102n1, 119, 121, 138, 156, 159, 163; and plantations 34; *see also* ecosystem services
bio-economy 139–140
biofuel 38n5, 71, 76, 165–166; *see also* fuelwood
biomaterials 139–140
biophysical conditions; effect on social-ecological systems 57
boreal forests 28, 58, 166
Borlaug hypothesis 156–156
Boserup theory 176–177; vs Malthus 177, 180
bounded rationality 87
Brazil 28–29, 31, 32, 57
Bruntland Commission 137
building *see* construction
bureaucracy, forest management 22, 50, 120, 138

Cambodia 66
Canada 27, 28, 66, 90
capitalism 22, 23, 25, 26, 65, 175
carbon 18, 36, 65, 87, 90, 100, 102n1, 139, 166; *see also* CO_2 fertilisation; *see also* greenhouse gas emissions
certification, forestry 26, 37, 65, 121, 156
change, the term 139
Chile 29, 159
China 20, 28, 32–33, 66, 80, 159
cities, and wood 20, 22, 24, 36
clearfell 160

climate change: carbon 18; and cultivated wood 35; fuelwood consumption 71, 76; historic 18–19; in New Zealand forestry 139
CO_2 fertilisation 165–166
coconut 131
colonisation 17, 21–22, 44, 53–55, 107, 110, 112, 117–118, 175
common pool resources 111–113, 123
community governance 111–112, 116–117, 120; *see also* self governance
competition, land use 18, 34, 166, 179–180
complex systems 2–3, 7, 46–48, 52, 85, 120, 174, 176–177, 179–180, 184
composite timber products *see* engineered wood
concession, forestry 33, 112–113, 116, 119, 121, 124
conflict 2, 3–4, 36, 52, 79, 86, 138, 160, 181; adding costs to natural forest wood harvest 37; forest loss 50; over plantations 33–34; policy tensions 101–102
conservation 16, 50, 55, 79–80, 81n5, 85–87, 89–92, 99–102, 102n1, 121, 138, 155–160, 167, 185; from plantation 119
conservation movement 52, 119, 138, 143
conservation reserves *see* protected areas
conservation value 89–91, 155; *see also* non-wood values
conservatism 142, 162
construction, and wood 20, 38n5, 92, 94
consumption 81n2; resources 23; wood and wood products 23–24, 48, 73, 76, 94–95, *96*, 100, 103n6
content analysis 128–129
contested resources *see* conflict
conversion, natural forest to plantation 33–34, 80, 97, 103n12, 159; *see also* deforestation
core, periphery (geopolitical) 23, 55, 107, 109
corporate ownership 33, 114, 160, 163
corruption 117–118, 121
crisis, in systems/society 47, 167, 180–181, 183
crop, trees 143–144
cultivated wood 1, 21, 27–35, 79–80, 128, 153–154, *153*, 163; definition 5–6, *5*, 8n1, 67; emergence 25; production estimates 71–74; productivity increases 29–30, 32, 51
cultural values 102n1; *see also* non-wood values
culture/nature *see* nature/culture
customary forest practices (ownership) 22, 108, 110, 112–113, 116–119, 121, 160, 163
cut 142–144, *136*

data: plantations 65; collection 164–165
decentralisation 107, 108–111; control of resources 115–116; (dis)advantages 111; international pressure 115; recentralisation 109, 116, 121; types of 110
definitional issues, forests and data 165
deforestation 18, 24, 28, 158; Indonesia 110, 115; plantation establishment 34, 80, 103n12; *see also* agriculture
development assistance 54, 55, 110–111, 115, 141
disciplines 48; multi-disciplinary 2–3, 46, 176, 180
discourse analysis 7, 128–129
discourse coalition 47, 182
displacement *see* leakage
distributional coalitions 47, 57, 173, 181–182
dynamic systems *see* systemic evolution

ecological impulse 25
ecological forestry 36, 79, 156, 163; *see also* sustainable forest management
ecological thinking 55
ecomodernism 179
economics: disruption 51; panarchy conditions 54; performativity of 146
ecosystem services 26, 34–38, 79–80, 102n1; in plantations 159; *see also* payment for ecosystem services
efficiency *see* intensification, of wood production; wood processing
employment 164; *see also* labour
endowment forest 15–16, 19, 27–28, 160–161
engineered wood 90, 140, 157, 176
England 28
environment 138–139; *see also* non-wood values; conservation
environmental awareness 139
equilibrium: in forestry 1, 52, 55, 129, 161; in systems 46, 162, 174; *see also* system evolution
eucalypt 31, 38n6, 54, 58, 100, 161

Europe 90; differences in wood/forest system evolution 55, 162, 174–175; early forestry 20–23, 28–29, 54; forests 90, 161
exotic species 28, 33, 128
export *see* trade
extant forests 20, 25, 27, 29, 50, 55, 154, 161, 174–176; *see also* endowment forest
externalities 87

FAO, forestry data 68–70
Faustmann 25
federal government 110
feedback: in policy 87; in systems 3, 47, 51, 184
fibre 1, 32, 90, 140, 161–163, 166, 176
fire 4, 18, 160; in cities 24
fisheries 47
food security 18, 35
forecasting 164
foreign aid *see* development assistance
forest: changing extent 50; definitional challenges 4–6, 53; fear of 35; valuing 35–36; *see also* non-wood values
forest ecosystem management 6, 16, 32, 35–38, 49–53, *49*, 122, 128, 131, 137, 155, 162, 163, 167–168; separating from wood production 6, 37–38, 85, 102, 139, 154, 185
forest ecosystem: definition 5
forest exploitation 6–7, 16, 19–21, 24–25, 49–50, *49*, 57, 108, 114, 122, 128–129, 144, 154; change to wood cultivation 20; stewardship forestry as 160–161
Forest Service: Indonesia 112–113; New Zealand 141–142
forest transition theory 38n2, 177–178
forester 52–53, 128–147, 155–156, 163, 168, 182, 185; defending profession 131, 137; definitional issues 53; emergence of 21–22; relevance 53; the term *forester* 131
Forestry 131–137; as territorial control 112; definition 6, 53, 146–147; history Indonesia 112–117; history of 16; institutions of 6
Forestry Law, Indonesia 112–113, 115–166, 118–119
fossil fuel 15, 22–24, 37, 166, 179
free riding 111
frontier and sovereignty 21–22
frontier exploitation 15–16, 21–22, 26, 49–50, 108, 142, 180–181
fuelwood 19, 21–23, 31, 69–74, 76, 81n4, 155
furniture 32, 92, 98–100, *98*, 103n11
futures: predicting 164–165; imagining 185

gatherer-hunter 16, 18–20, 38n1
gaming, institutional 165
GDP, contribution by forestry 164
Germany, forestry origins in 21–22
global: analysis 65–80; rationale for 65; wood supply 156
globalisation 58, 107, 141, 166
governance 47, 107; forestry 115, 121–122
grass crops 166
greenhouse gas emissions: in leakage 87

hardwood 33, 91–92, *96*, 97–99, 103n11, 113; plantations 86
harvest 142–144, *137*
high grading 160
history, and panarchy 54, 174–175
history, thematic 15–17, *17*
housing 94; *see also* construction
Hubbert curve 66, 78, 80
hunter-gather *see* gatherer-hunter
hunting 21, 35, 119

illegal logging 108, 114, 122, 124
import *see* trade
India 22, 55
indigenous forest *see* natural forest; *see also* New Zealand
indigenous peoples 22, 108, 117
indigenous species 5, 34, 92
Indonesia 80, 107–127, 175; changes in wood exports 114–115; history of forestry 112–117; introduction 7–8; location 7; wood production 8, *9*
industrial roundwood 8, *30*, 60, 69–76, **69**, **72**, *74*, *75*, *76*, 103n5, 153; definition 67; from plantations 29, *30*, 81n3
industrialisation 22–25, 54, 107–108, 122, 173, 177, 178
industry, timber and wood processing 23–24, 30, 32, 52, 57, 113–114, 128, 139–141, 158, 160, 163–164
inertia in resource use social systems 56, 162, 167, 173, 180–185
innovation 176–180; induced innovation 23–24, 51, 56, 100, 158–159, 177; redefining 177–179; social 179

institutions 3, 47, 122, 146; definition 10n3; forestry 16, 21, 25–26, 47, 49–56, 108–109, 118–119, 128, 131, 146, 162, 164, 168, 179, 185; institutional bias 1; weak 122; *see also* forester
integrated land use *see* land use
intensification, of wood production 29, 31, 34, 50–51, 86, 101–102, 119–120, 128, 131, 147, 153, 155–157, 176–177
interdisciplinary *see* disciplines
international forest policy 115
IPAT model 178–179

Japan 20, 57, 97
Java 1, 28, 54, 109, 112–113, 118, 122n1, 123n5, 123n6
Jevons paradox 158

Kalimantan 108, 112, 113, 116
knowledge 57, 146, 184; knowledge transfer 48, 55–56, 141; limited 87
Kuznets curve 177–178

labour 164; costs 24
land clearing 15–16, 18–20, 108
land competition 18–19, 166
land management 52–53
land sharing 8, 155–156, 184
land sparing 100–102, 155–156, 184
land tenure 37, 50; in Australia 85, 91; in Indonesia 108, 116, 118; *see also* private forest; protected area
land use intensification *see* intensification
land use: land use integration 156, 162, 167; land use policy 85–91, 101–102, 113–116, 157–159, 164–168; patterns 49–50, 159
landscape approach 167, 180, 185
Laos 66
Latin America 31, 32
leakage 85–102, *88–89*, 157–159, 178
leviathan governance 108, 111, 113, 118–119
license, logging *see* concession, forestry
locking-up, forest through conservation 156
lock-in 31, 56–57
localised ownership 163
log 142–144, *136*
log production *see* wood production
logging *see* wood harvest
logic: of wood and forest social ecological systems 6, 16, 22, 36, 50–51, 53, 55–57, 137, 154, 178, 185; of decentralisation 120
long production cycles, wood 161, 162

Malaysia 27, 32, 66
Malthus 176–177
management 137–138, *131*
manufacturing 23, 164; *see also* wood processing, furniture and
Maori 18
mapping 117–120, 123n13
market 90, 103n7, 111, 140, 146, 161, 179; decentralisation 110; land 158; plantation wood 79, 101; wood products 31, 38n5
mean annual increment 31, 38n10
mechanistic systems, or thinking 48
modelling 80, 88–91, 99, 158, 164, 174; global roundwood 66–80
modernity 3–4, 21, 122, 141–142, 154, 173, 175, 180; *see also* reflexive modernity
monoculture 21, 31, 143, 155–156
multidisciplinary *see* disciplines
Multiple Use Sustained Yield Act 26
multiple values, in forests *see* non-wood values
multiple-use forestry 16, 25–27, 36, 52, 156, 163
Myanmar 66

nation states: centralising tendency 117; forestry 21–22; sovereignty 32; unit of study 65
national park *see* protected area
native forest 102n2; *see also* natural forest
natural forest: costs of harvest 29; definitions 4–6, 5, 66–68; global wood production 77–78; history of human impact 18–19; non-renewable 78–79; peak production 78
nature 180
nature/culture 4–5
neoliberal 185
New Order, Indonesian 107, 110, 113, 115, 117–118, 121
New South Wales 26, 92, 97, 100, 103n11
New Zealand 28, 29, 128–147, 159, 161, 175, 176; distinctive forest 57; early deforestation 18; introduction 7–8; location 7; logging ban 79, 139; wood production 8, *9*

New Zealand Institute of Foresters 128, 136
New Zealand Journal of Forestry 128, 130
non-wood values 25–27, 34–38, 49–51, 54, 80, 86, 102n2, 131, 136, 143, 151, 156, 174, 182, 185; impact from wood harvest 157; impact on wood harvest 140; social drivers of 52
normative 146, 155, 166
North America 30, 90, 166

old growth forest 29; *see also* primary forest
optimising, land use or production *see* intensification
Outer Islands, Indonesia 54, 112–113, 120; definition 123n6

Pacific North West 27, 29
palm oil 31, 114; within forestry 53; wood source 31–32
panarchy 47–48, 52, 53–58, 145; definition 47; historical influence 174; local distinctiveness 122, 175; technology 55
paper *see* pulp and paper
Papua New Guinea 66
park 138, 144; *see also* protected area
path dependence 3, 56, 174
payment for ecosystem services 36–37, 163
peak wood 66, 78, 80, 153
per capita: wood consumption 24, 70, 76, 81n4, 95, 100, 103n6; wealth 24, 95, *95*
performativity 146
perpetuity 137
Philippines 66
pine species 38n6, 57–58, 91, 99, 101, 143, 145, 161; native Australian pine species 92
plague 21
plantation 27–31, 71–73, 81n6; definitions 4–6, *5*, 67–68, *68*; expansion (Indonesia) 114; global expansion 29, 73; global move to South 28–29; increased productivity *see* cultivated wood; fast growing; political drivers 32–33; national promotion, schemes 145; negative impacts 33–34, 79–80, 159; social impacts 79–80, 159; *see also* subsidies
plantation conservation benefit 37–38, 154, 156–160
planted forest 71–73, 81n6; definitions 4–6, *5*, 67–68, *68*
plywood 92, 114, *114*
polarisation 181–182
policy: integrated policy 159; *see also* land use; plantation in Australia 26–27
political 2, 4, 10n3, 37, 47, 55–57, 101, 107–120, 181–182; plantation expansion 32–33
poplar 31
population growth 16–19, 21, 23, 24, 50–51, 164, 176–179
Porter hypothesis 177
post-forestry 16–17, *17*, 27–38, 53, 85, 147, 167–168, 173, 175, 185
power, political 108; over forests 110; contestation 117
preservation *see* conservation; *see also* protected area
primary forest 160
prisoner's dilemma 111
private forest 50, 90–91, *95*, 96–97, 100, 103n7, 111, 158, 163
privatisation 110–111, 113, 118
production forest 26, 85, 91, 143, 162
productivity: in agriculture 32; in cultivated wood 32, 34, 38n9, 51, 58, 101–102, 159–160, 179; in natural forests 28, 51; in wood processing 90, 100, 179
property: and stewardship forestry 50; *see also* customary forest practices (ownership)
protected area management, managers *see* land management; *see also* forest ecosystem management
protected areas 27, 37, 50, 86, 147, 156
province *see* regional government
pulp and paper 23–24, 30–32, 56, 114, *114*, 140
pulplog 67, 94–98, *94–96*, *98*, 100, 103n4, 110, *114*

Queensland 26, 92

rainforest *see* tropical forest
rebound effects 158
reconstituted timber products 176; *see also* engineered wood
recreation 36, 50, 91, 177
REDD 119–120
reflexive modernity 3–4, 183–185
reforestation 20, 28, 33, 90

Regional Forest Agreement 86–87, 91, 93–94, 98–99, 101
regional government 8, 26, 57, 65, 110, 112–113, 116–118
regulation 21, 26, 47, 50, 112–113, 115–116, 120–121, 158–159; environmental 51, 163, 177; deregulation 110
renewable resources 78
research and development 140
resilience in resource use systems 3, 46–58, 122, 162, 167–168, 173–176; from cooperation 119–120
resilience thinking 46–48; definition 3; review and considerations for application 173–176
resistance to change 183; *see also* inertia in resource use social systems; *see also* resilience in systems
resource use systems 2, 8, 154, 168, 173, 179–181
restoration ecology 145–146
Rio Earth Summit 137
risk society 3, 183–184
roundwood: definition 67; global production 68–78; global projections 70–71
rubber 31–32, 53
Russia 28, 66, 90

sawlog 67, 92–101, *94–96*, *98*, 103n4, 103n11
science: changing 26; facts 182, 184; of forestry 25, 145, 184; limitations of 184; *see also* social science
secondary wood products 92
selective logging 58, 160; in Germany 163
self sufficiency, in wood 32, 56–57
semantics, *forests*, *forestry and foresters* 53
separation, of wood and forest 162
shifting agriculture 15, 18, 33, 38n3
shipping, wood trade 22–23
silviculture 34, 55, 71, 138, 145, 160–161, 168n3
social forestry 119–120, 160
social science 3, 48, 184; natural system analogy 48; *see also* science
social values *see* values
social-ecological systems 165, 173–185; failsafe 183; 'flip' between states 122; introduction 3; in modernity 141; ship analogy 181–182; social inertia 181; *see also* wood and forest social-ecological systems
Soeharto 107, 108, 113, 115, 118, 121, 123n3
Soekarno 112–113
softwood 86, 91–92, *96*, 97
soil and plantations 33–34
Solomon Islands 66
South Africa 28–29, 73
South Australia 8
sovereignty *see* nation states
specificity test 130–131; results *132–136*
spheres of interest 108–112, *109*, 119–122, 122n2; in leakage policy 87
stability: of system 51, 162, 174; in forestry 50, 56
standing stocks *see* extant forests
statistics *see* data
stewardship forestry 6, 16–17, 17, 20–27, 49, *49*, 108, 122, 160–168, 182–183; co-opted by colonisation/capitalism 22; defined 20; developed vs developing countries 122; failure to deliver aims 27; religious quality 52; sustainability 178
stressors 47
subjectivity 182
subsidies 32–33, 57, 114, 157; MIS, in Australia
substitution, wood and wood products 90, 97–102, 179
Sumatra 33, 108
sunk costs 47; *see also* path dependence
supply chain 100
sustainabilities, multiple 183
sustainability 1; avoiding overshoot 179–181; bounded 22; territorial control 22; contested 180–185; forestry 17, 20–27, 49, 51–52, 129, 154; optimal 183; technology/innovation 176–180; the terms *sustainability* and *sustainable* 137–138, 144; wood production 80
sustainable forest management 16, 79–80
sustainable yield 23, 25, 52, 141, 158, 160–161; in Australia 90–94, 100, 103n4, 103n5
swidden *see* shifting agriculture
system evolution 1

Tasmania 8, 18, 182
teak 28, 54, 112
technology 24, 30, 38n11, 51, 54–56, 129, 139–141, 164, 167, 176–180, 3D printing 176; and panarchy 55; *see also* induced innovation *and see also* wood processing

technology transfer 141; *see also* development assistance
temperate forests 28, 38n9, 58, 166
Thailand 66, 79
three tacks of analysis 7
timber harvest *see* wood harvest
trade 20, 54, 65, 85, 92, 98–100, *98*, 157, 179; Indonesia 114
tradition 141–142
traditional forest practices *see* customary land practices
tragedy of the commons 50, 108, 111, 119
transaction costs 163; definition 38n12; conservation in wood harvest 37
transformation 46–48, 52–54, 58, 154, 162–163, 167–168, 176; vs adaptation 52–53, 167–168
transition in wood sourcing 1–3, 6–8, 36, 65, 79, 85, 102, 118, 122, 128–129; between social-ecological system basins 144–145; as land sparing 155; messy 144; *see also* forest transition theory
tree as crops 53; *see also* wood cultivation
trees outside forests 31–32, 71–74
triangulation 7
tropical forests 18, 57–58, 66, 90

United States *or* US 25, 27, 28, 29, 30, 52, 55, 57, 73; forestry origins 22
unsustainable wood production 20, 27, 50, 57, 66, 79, 114–115, 139, 153–154
unsustainable resource use 179–182

values 51, 183; *see also* non-wood values
Vietnam 32–33, 79, 157, 178
Von Thunen 15, 38n2

water: limiting wood cultivation 165–166; impact of plantations 33
Western Australia 97
whole log 99; in Indonesia 113–114, *114*; whole log equivalents 92, 94–95, 99
wilderness 27
willow 31
wood and forest social-ecological system 3–4, 7–8, 47–49, *49*, 51–58, 85, 107–108, 122, 128–129, 138–139, 142–146, 154–155, 162–168, 173–185; introduction 3
wood chip 33, 97
wood cultivation 27–35, 37–38, 108, 128–129, 139, 143–145, 154–155, 157, 166; to address conservation needs 26–27; agricultural 6, 46, 66; definition 6–7; global production 67–78; historical examples 28; late start 20
wood cultivation, as basin 48–49, *49*, 51, 57–58, 154–155, 165–166, 177–180, 185
wood fuel *see* fuelwood
wood harvest: ban 113; impact on non-wood values 157; reducing harvest 23–24; restrictions 177; reduced intensity 27; separating from forest management 37–38; to pay for conservation 36, 163
wood processing 114; alignment with plantation wood 30–31; capital intensive 30–31, 51, 114, 120–121; increasing efficiency 30–31; reducing costs of 23–24; technological advances in 23–24
wood products 23–24, 30, 92, 114, 140; environmental advantages of 37; in multi-storey buildings 38n5; *see also* wood processing; secondary wood products
wood shortage 19–24
wood substitution *see* substitution
wood/forest nexus 1, 6, 16, 36–37, 52–53, 143, 146–147, 165, 168, 173
world system 20, 48, 54, 56